Beyond Climate: Full Pollution Fight

Chris

Table of Contents

Chapter 1: Introduction

1.1   Background and Motivations

Human beings produce vast quantities of pollution taking many forms and having deleterious effects across earth's natural cycles and systems. Climate change driven by greenhouse gas emissions threatens devastating changes throughout the biosphere, and has therefore come to the forefront of scientific and regulatory attention toward pollution. However, humanity's activities produce many other forms of pollution that threaten our natural environment, and in turn our way of life. Environmental damage from a variety of chemical emissions and wastes beyond greenhouses gases is far-reaching. Nitrogenous and porphyritic by-products of agricultural and urban activities, as well as the combustion of fossil fuels, have altered the nitrogen and phosphorus cycles. Known effects of the disruption of these cycles includes the acidification of soils, reduced bioavailability of nutrients for the support of flora, nutrient flooding of surface waters causing toxic algae blooms, reduced dissolved oxygen in aquatic environments, loss of coral reefs, and reductions in the biodiversity of aquatic ecosystems [1], [2]. The use and disposal of plastics has led to microplastic contamination of marine environments, the scope and effects of which are not yet understood [3]. Heavy metal waste from industrial processes is growing, causing toxicity in plants and bioaccumulation in animals, including species relied upon by humans for food [4]. Other pollutants are non-chemical in nature: notable examples are light and noise. The advent of electrical lights has led to a temporal reorganization of human activities and light availability in human-influenced environments. Animal breeding timing and behaviours, migration patterns, and predator-prey relationships have all been observed to be affected by light pollution [5]. Noise pollution, the focus of the present work, is of particular concern in underwater environments, as aquatic fauna tend to rely heavily on sound to perform basic life functions. Underwater noise pollution is known to adversely affect the communication, sensing, feeding, stress levels, and mating behaviours of marine animals through auditory masking and hearing damage [6], [7]. This **book** examines radiated noise from marine vessels as a damaging pollutant; its primary aim is to promote expanded capabilities in predicting propeller noise to facilitate the mitigation of that noise for the benefit of marine environments.

Sound is transmitted much more efficiently in water than in air. Many marine animals have adapted to take advantage of their aquatic environment by relying on sound as a primary sense. Negative impacts from anthropogenic noise have been observed in many different aquatic animals, notably multiple species of cetacean (e.g. [8]–[11]) and fish (e.g. [12]–[14]). The consequences of acute exposure to anthropogenic noise can be severe; stranding and death have been recorded in beaked whales as a result of exposure to sonar signals [11]. Chronic effects from noise pollution have been difficult to monitor, but there is cause for concern. Reductions in the effective communication ranges of multiple whale species due to increased background noise levels associated with human activity has been noted in the literature [6], and long term exposure to anthropogenic noise is thought to be a contributing factor in the decline endangered whale populations [9], [15], [16]. In addition, the efficient transmission of sound also means that the range of effect of a single source of noise pollution can be on the order of hundreds or thousands of kilometres [7].

Recent understanding of the effects of anthropogenic underwater noise on marine ecosystems has led to increased scrutiny of the underwater radiated noise from ships. Ships represent the largest source of anthropogenic underwater noise in, and a growing number of vessels has coincided with a continuous and decades-long rise in ambient noise levels throughout the world's oceans [17]. Under typical conditions, underwater radiated noise from ships is dominated by cavitation-induced noise from propellers [18]. As water flows around propeller blades, the reduction in local pressure can cause vapour cavities to form within the liquid medium. The oscillations and eventual collapse of these vapour cavities are associated with high-energy acoustic emissions. It is crucial to develop the ability to predict propeller cavitation and the noise it generates in order to mitigate the underwater noise pollution, and propeller cavitation in has become an area of increased research interest as a result of the increased attention to underwater noise. Analytical models to determine acoustic noise from perfectly spherical bubbles have existed for many years [19]. In reality, however, cavitation bubbles do to not exist as perfect spheres. Instead they have a wide range of shapes, sizes, and often contain non-uniform regions of both liquid and vapour. As a result, solutions are not readily obtained from analytical methods.

Experiments have proven more useful to study and characterize cavitating flow properties. Numerical techniques have also been used to predict cavitation structures around propellers, and provide the advantage of addressing the lack of detail and expense of conducting experiments. Simple potential flow solvers have been effective for predicting cavity structures that are attached to solid surfaces [20]–[22], while Unsteady Reynolds-Averaged Navier-Stokes (URANS) solutions have shown promise for the prediction of cavitating vortices [23]–[27]. However, higher-order simulations would be required to directly predict acoustic emissions, but come at significantly increased computational expense [28]. To predict acoustic output, recent studies have proposed using semi-empirical noise models based on numerically-predicted cavitation structures and experimental measurements of noise under controlled conditions [29], [30].

This **book** aims to expand upon existing numerical and empirical modelling strategies to evaluating cavitation noise through their application to ship propellers, and to provide insight into future directions that these methods may take. Chapter 2 concerns a combined experimental and numerical (using URANS solutions) study of noise from a model-scale cavitating propeller. Chapter 3 outlines a procedure for mapping cavitation noise on an engine-parameter space, a technique targeted toward application in noise optimization routines, and gives a proof-of-concept using a panel method simulation code to predict noise from the model propeller discussed in Chapter 2. Chapter 4 takes the reverse approach to modelling and analysing cavitation noise; field measurements of radiated noise from coastal ferry vessels are analysed in an attempt to relate the features of the noise signatures to simple operating conditions of those vessels.

The present **book** is structured as a compilation of manuscripts and paper drafts that have either been submitted or that are intended to be submitted for archival journal publication. Each of Chapters 2, 3, and 4 are written as standalone papers. Some modifications, such a formatting changes, have been implemented for the sake of consistency.

1.2 Cavitation and Cavitation-Induced Noise

1.2.1 The mechanism of cavitation

Cavitation is a phase-change process in which material transitions from a liquid phase into
a vapour phase. Although cavitation is closely related to the more familiar process of
boiling, the two are distinct in their thermodynamic paths. In boiling, a liquid ruptures to
form vapour as a result of an increase in temperature; in cavitation, rupture is induced by
a reduction in pressure. The basic principal of cavitation can be illustrated on a simple p-
V (pressure-volume) phase diagram such as the one shown in Figure 1.1. In the simplest
case, assuming ample nucleation for bubble formation and a slow process, cavitation may
be described by the process denoted by curve A-B-C-E. When the pressure of the liquid
drops to the vapour pressure at point B the phase change begins, following the horizontal
isotherm through the two-phase region toward point C as the volume increases. If the
pressure continues to fall to point E after the volume of liquid has transitioned entirely
into a vapour, the process will continue into the gaseous region.

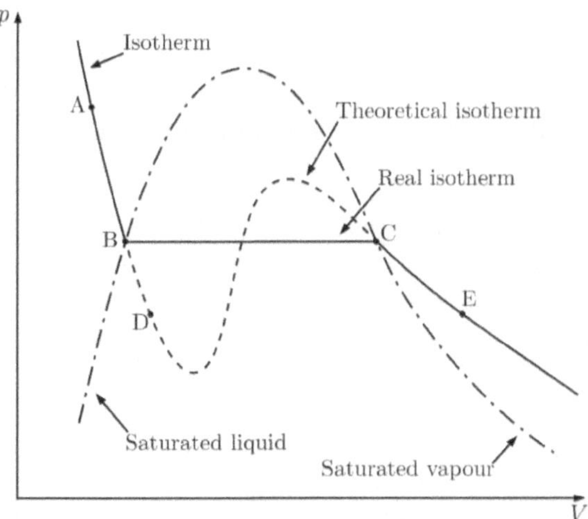

Figure 1.1: A p-V phase diagram depicting the liquid-vapour two-phase region.

Nucleation sites promoting the formation of bubbles are limited in real liquids, and as
a result, liquids can typically withstand some tension before rupture occurs. Point D in

Figure 1.1 depicts a state of tension; the magnitude of that tension is the difference between the pressure at point D and the vapour pressure (the pressure at point B). In the absence of sufficient nucleation, the depressurization process A-E would follow the theoretical isobar to the metastable state D rather than following the horizontal isobar through the two-phase region. A small disturbance to state D would result then in a transition to state E.

The amount of tension a liquid can withstand before rupture and the onset of cavitation is known as the tensile strength of the liquid, and the magnitude of the tensile strength determines the critical pressure for cavitation inception. The bulk tensile strength of a liquid is dictated by points of weakness within the medium. These weaknesses result from material inclusions in most engineering applications; however, homogeneous nucleation theory claims that microscopic vapour inclusions occur naturally and provide nucleation sites for rupture events even in homogeneous liquids. Solid surfaces present in liquids tend to have microscopic imperfections with geometric conditions that result in a local tensile strength close to zero, and the inception of macroscopic cavitation events therefore tends to occur on these surfaces [19]. In the context of marine propellers, the concentration of nucleation sites on the surface of propeller blades typically results in cavity formation on the propeller surface at pressures close to the vapour pressure of seawater.

1.2.2 Spherical bubble motion and noise

Vapour bubbles from cavitation produce sound through pulsation and collapse. Assessment of near-field noise from bubbles requires acoustic solutions to the equations of bubble motion. Analytical descriptions of bubble motion are, in most cases, known only for spherical bubbles. In the context of cavitation, many modelling strategies are based on the Rayleigh-Plesset equation, which describes temporal evolution of the radius R of a spherical, adiabatic bubble filled with saturated vapour and gas in an unbounded Newtonian liquid:

$$\rho \left[R(t) \frac{d^2 R(t)}{dt^2} + \frac{3}{2} \left(\frac{dR(t)}{dt} \right)^2 \right] = p_v - p_\infty(t) + p_{g0} \left(\frac{R(t)}{R_0} \right)^{3\gamma} - \frac{2\gamma_s}{R(t)} - 4\rho v \frac{1}{R(t)} \frac{dR(t)}{dt} \, , \quad (1.1)$$

where ρ is the density of the liquid, $R(t)$ is the bubble radius, p_v is the vapour pressure of the liquid, $p_\infty(t)$ is the ambient pressure, p_{g0} is the equilibrium pressure of the gas, R_0 is the equilibrium radius of the bubble, γ is the ratio of specific heats of the gas, γ_s is the surface tension of the bubble envelope, v is the kinematic viscosity of the liquid, and t is time. The Rayleigh-Plesset equation plainly illustrates the dependence of the evolution of a vapour bubble's radius on the difference between the vapour and ambient pressures, the equilibrium gas partial pressure and bubble radius, as well as the surface tension and liquid viscosity. Neglecting the viscosity term, which is justified at scales where inertial forces dominate, results in the Rayleigh equation. Both the Rayleigh and Rayleigh-Plesset equations are widely used in bubble modelling, with the Rayleigh equation favoured for modelling bubble collapse. These equations lose validity in the final stages of bubble collapse, where liquid compressibility becomes relevant, but the Rayleigh equation is nonetheless effective in capturing the short duration and rapid change in scales during bubble collapse [31]. Solution of the radius evolution facilitates approximate solution of the acoustic pressure in the far field, which depends on the second time derivative of the bubble's volume V:

$$p(r,t) - p_0 \cong \frac{\rho c}{4\pi r c_0}\left(t - \frac{r}{c_0}\right)\left(\frac{d^2 V}{dt^2}\right),$$ (1.2)

Where r is the radial distance from the bubble centre, $p(r,t) - p_0$ is the acoustic pressure at a distance r from the bubble, c is the local speed of sound, and c_0 is the equilibrium speed of sound. Radiated acoustic energy E can also be estimated from the volume evolution:

$$\frac{dE}{dt} \cong \frac{\rho}{4\pi c_0}\left(\frac{d^2 V}{dt^2}\right)^2.$$ (1.3)

Since solutions for the motion of non-spherical bubbles are not readily available, numerical solution schemes often approximate cavities of other shapes as collections of spherical bubbles. This approach is advantageous for its simplicity, but notably neglects proximity effects.

The use of Rayleigh-Plesset or Rayleigh equations also neglects thermal delay; in reality, heat must be supplied from the liquid for vaporization to take place, which requires time that can become important at small scales. For further information on thermal effects the reader is referred to [32].

1.2.3 Propeller-induced cavitation regimes

The structure of propellers and the flow about them tends to result in cavitation appearing in particular patterns. It is convenient to categorize these patterns as regimes. Almost all cavitation induced by propellers can be categorized as tip vortex, hub vortex, sheet, bubble, or cloud cavitation. The tendency for cavitation of all types to be induced by propellers is dependent on the nominal cavitation index,

$$\sigma_N = \frac{p_{ref} - p_v}{\frac{1}{2}\rho n^2 D^2}, \tag{1.4}$$

where n is the propeller revolution rate (in revolutions per second), D is the propeller diameter, p_v is the vapour pressure of seawater. The hydrostatic reference pressure is calculated according to: $p_{ref} = p_{atm} - \rho g z_0$, where p_{atm} is the atmospheric pressure, g is the gravitation acceleration, and z_0 is the subsurface depth of the propeller shaft. The inception of different types of propeller-induced cavitation tends to occur at different values of the cavitation index, with higher values of σ_N corresponding to a lower likelihood of cavitation.

Vortex cavities exist within the core of shed vortices from the propeller. The hub vortex coalesces from vortices shed at the root of individual blades and trails behind the centre of the propeller. No hub vortex cavitation was observed in the experiments discussed in Chapter 2, and discussion on cavitation of this type is therefore limited throughout this book. Tip vortices are shed from the tips of the blade, and tip vortex cavitation tends to occur at relatively high cavitation indices. While other forms of cavitation tend to be more common during off-design operation of a propeller, tip vortex cavitation is common during the normal operation of many vessels. At higher values of σ_N the vortex may be unattached, and as σ_N the cavity connects with the blade.

Sheet cavitation most commonly occurs on the suction side of propeller blades, close to the tip, and takes the form of a sheet covering the blade's surface. It is often connected to tip vortex cavities. Sheet cavitation typically occurs at smaller values of σ_N than tip vortex cavitation. As the cavitation index is reduced further, bubble cavitation, characterized by the formation of discrete, macroscopic bubbles on the surface of the propeller blades, may also occur. Cloud cavitation consists of smaller bubbles, and is generally shed into the flow from sheet or bubble cavities. When a propeller is operated in under-loaded conditions, sheet cavitation can also occur on the pressure side of blades.

Noise from propeller cavitation of all types is primarily broadband in nature, with the notable exception of singing tip vortex cavitation. Tip vortex cavities have a resonance frequency related to the diameter of the cavitating vortex core; excitation of the vortex core may result in resonance at a specific frequency known as singing [33], [34]. As a result, singing tip vortex cavitation can be identified in an acoustic signature by the existence a sharp peak.

Experimentally obtained images of different cavitation regimes and their associated acoustic signatures can be found in Chapter 2.

1.3   book Structure

The body of this book is divided into three main chapters, each of which is was written with the intention of submission to an archival journal as a standalone article. This subsection outlines this author's contributions to each paper, as well as those of each of the co-authors. Some modifications have been made to the formatting of these papers for the purpose of consistency, including the numbering of headings and referencing styles.

1.3.1 Contributions

Chapter 2

McIntyre, D., Rahimpour M., Dong, Z., Tani, G., Miglianti, F., Viviani, M., Oshkai, P. (2020). **"Measurements and numerical simulations of underwater radiated noise from a model-scale propeller in uniform inflow."** Submitted to Ocean Engineering Sep 21, 2020.

The first manuscript details a combined experimental and numerical study of noise from a cavitating propeller, and focuses on both the fundamental importance of experimental findings and the effectiveness of the numerical modelling strategy used. Experiments were run, and their results post-processed at the University of Genoa by Drs. Tani and Viviani and Ms. Miglianti. Mr. Rahimpour at the University of Victoria designed and submitted the cases for computation at performance computing facilities with Compute Canada. I post-processed the simulation solutions and performed the acoustic analysis. I also performed the comparison between the simulations and experiments and the related analysis. Drs. Oshkai and Dong provided guidance in terms of research direction and supervision. I wrote the original draft of this manuscript, and the present version has been edited with the assistance of co-authors.

Chapter 3

McIntyre, D., Dong, Z., Tani, G., Miglianti, F., Viviani, M., Liu, P., Oshkai, P. (2020). **"Parametric mapping approach for cavitation-induced acoustic emissions from marine propellers."** In preparation for submission to Ocean Engineering.

In the second manuscript I present a methodology I developed for generating "maps" relating cavitation noise to the speed and torque of a ship's propeller. These maps are intended to provide a framework for predicting noise pollution by borrowing the concept of emissions mapping used to predict chemical emissions from internal combustion engines in the context of optimization. This paper contains references to the methods and results presented in the paper comprising Chapter 2; external references to that paper have been replaced by internal references for clarity. Given the central role of those experiments to the results presented in this work, the experimenters (Tani, Miglianti, and Viviani) have been included as co-authors. Figures containing schematics and tables of experimental conditions shared between both papers have been excluded from Chapter 3, and their Chapter 2 versions are references instead. ROTORYSICS, the numerical code used for simulations in this manuscript, is the work of Dr. Liu of Newcastle University. Drs. Oshkai and Dong provided guidance in terms of research direction and supervision. I performed the numerical simulations, noise model formulations, and data analysis.

Chapter 4

McIntyre, D., Lee, W., Frouin-Mouy, H., Hannay, D., Oshkai, P. (2020). **"Influence of propellers and operating conditions on underwater radiated noise from coastal ferry vessels."** Submitted to Ocean Engineering Sep 23, 2020.

The final manuscript is an analysis of field noise measurements of coastal ferries in commercial operation provided by JASCO Applied Sciences Ltd. and BC Ferries Corp. The original analysis of this data was performed by Dr. Frouin-Mouy of JASCO. The portions of the analysis that discuss narrow-band radiated noise level (RNL) trends (section 4.5.1), multivariable linear regressions (section 4.5.2), and radiated noise regime analysis (section 4.5.3) are my original work. The discussion of narrow-band spectral features (section 4.5.4) is the work of Mr. Lee. I wrote the original draft of the present manuscript, with the exception of section 4.5.4 (written by Mr. Lee). Edits to the manuscript have been made by my co-authors. Guidance in terms of research direction was provided by Mr. Hannay of JASCO and Dr. Oshkai.

Chapter 2: Experimental Measurements and Numerical Simulations of Underwater Radiated Noise from a Model-Scale Propeller in Uniform Inflow

Duncan McIntyre [a], Mostafa Rahimpour [a], Zuomin Dong [a], Giorgio Tani [b], Fabiana Miglianti [b], Michele Viviani [b], Peter Oshkai [a]

[a] Department of Mechanical Engineering, University of Victoria, Canada

[b] Department of Naval, Electrical, Electronic and Telecommunications Engineering, University of Genoa, Italy

2.2 Keywords

Propeller-induced cavitation; underwater radiated noise; cavitation-induced noise; RANS; CFD

2.3 Introduction

Ambient noise levels are a critical measure of the health of marine ecosystems, where fauna favour sound as a means of communication and sensing [6], [8], [10]. That health is threatened by noise pollution from human activity, of which shipping is the largest source [17]. Shipping noise can be classified into two categories related to its source. Cavitation, when it occurs, is the dominant source of noise underwater noise from ships [18]. Cavitation phenomenon, i.e. formation of vapour bubbles in the regions of the liquid flow field where local pressure drops below a critical level, is induced primary by propellers and is strongly influenced by operating conditions of the vessel.

Reliable prediction of propeller-induced cavitation noise has been the topic of increased researched interest in recent years. The numerical techniques for prediction of propeller cavitation included applications of boundary element methods that resulted in accurate prediction of cavitation-induced loading on marine propellers [35]. More recently, several research groups successfully applied techniques based on the solutions of unsteady Reynolds-Averaged Navier-Stokes (URANS) equations and volume-of-fluid treatment of the two-phase flow region to cavitating flows around marine propellers, enabling prediction of the total cavitation volume as well as cavities induced by tip vortices [23]–[25], [27], [36], [37]. Large Eddy Simulations (LES) of a cavitating propeller has also been undertaken (e.g., [38]), although relatively high computational expense of this methodology has so far limited its adoption in the field.

The numerical prediction of shipping noise has also progressed with the advances in available computational resources and techniques. However, availability of acoustic data in civilian applications has historically been limited. To the best of our knowledge, the first direct comparison between numerical simulations and full-scale measurements of radiated noise from ships, which neglected cavitation-induced noise, was presented by Ianniello et al. [39]. The same authors later presented a methodology for numerically predicting cavitation-induced noise (CIN) of a marine propeller based on the solution of URANS equations and Detached Eddy Simulation (DES) as inputs of a modified form of the Ffowcs Williams-Hawkings (FWH) acoustic analogy equation [40]. A more recent study by Wu et. al. adopted a similar procedure to reproduce a set of model scale experiments, comparing hydrodynamic parameters between the experiments and numerical solutions [41]. A small number of studies have acoustic measurements from scale model tests of cavitating propellers to computational fluid dynamics (CFD) results. Kowalczyk and Felicjancik performed a comparative study with a scale model propeller and URANS simulations, with acoustic comparison limited to a single loading condition; results showed good agreement at low frequencies but diverged at frequencies above 100 Hz [42]. Li et al. compared full scale sea trial acoustic data with scale model experiments and numerical simulations using DES and the FWH analogy; both the numerical and scale models under-predicted broadband noise levels at the majority of frequencies [43]. Sezen et al. recently presented a URANS-based acoustic benchmark of a test propeller and found good agreement for low-frequency noise [44]. Despite the valuable insight provided by existing numerical studies, there is a need for further direct comparison between experimental and numerical acoustic data from cavitation propellers in the literature.

Experimental studies of cavitation-induced noise from marine propellers have been conducted at both model and full scales in literature, most commonly using dedicated cavitation tunnels. One study taking advantage of both a cavitation tunnel and research vessel to provide a comparison of model-scale and full-scale radiated noise data, noting both the success of the validation methodology and the high degree of technical challenge involved in performing a study of its kind [45]. A similar study was performed by Tani et al. with the same propeller examined in the present work [46]. Several studies have examined tip vortex cavitation noise specifically, motivated by its relative prevalence in normal ship operation. One study was able to compare acoustic measurements of tip vortex cavitation-induced noise from scale model and full-scale propellers in order to determine a scaling exponent [47], which was found to agree with previous theoretical work by Shen, Gowing, and Jessup [48]. Another study was able to determine that the dominant oscillation frequency of the tip vortex cavity was related to the zero group-velocity condition in the cavity's "breathing" mode [33]. All three of the aforementioned studies note the significant influence of vortex dynamics, something not often reproduced well by URANS simulations of propellers, on sound generation. Other experimental works have considered acoustic design optimization using experimental methods. A study using experiments to validate panel method code for noise optimization of a controllable-pitch propeller (CPP) showed good reproduction of cavity extent by the code, which was correlated with radiated noise in the experiments [49]. Other work still has looked at the acoustic signatures related to specific cavitation regimes, e.g., [50].

The mechanism of sound radiation from macro-scale propeller-induced cavitation structures is not well understood, and the numerical simulation of acoustic noise generated by a cavitating propeller is complicated by the need to model both the cavitation phenomenon at appropriate time and length scales as well as the larger hydrodynamic field around the propeller. The present work examines the cavitating propeller noise source and its numerical simulation by combining model-scale experiments with CFD solutions.

2.4 Methodology

2.4.1 Overview

The present study uses the methodology presented by Ianniello and De Bernardis [40] in combination with a set of model scale laboratory experiments conducted in the cavitation tunnel of the University of Genoa. We measured both hydrodynamic and acoustic quantities during the experimental campaign. The entire test section was reproduced in the numerical domain, which allowed for direct comparison between the experimental hydrophone measurements and numerical data at the location of the hydrophones. Uniform inflow was considered instead of a more typical simulated wake inflow condition in order to study cavitation noise in the most simplified case by isolating it from intermittency in the flow field. A range of operating parameters was examined to produce a representative set of cavitation phenomena. The cavitation types observed in the current study were tip vortex cavitation, sheet cavitation, pressure-side cavitation and bubble cavitation. The corresponding cavitation patterns are shown schematically in Figure 2.1. Control of the cavitation number and thrust coefficient allowed the noise contributions from these individual types of cavitation to be studied both in isolation and in various combinations. Quantitative comparisons between experimental measurements and simulations were made in terms of propeller thrust, torque and acoustic power spectra. The quantitative **analysis** was supplemented with qualitative comparisons of numerically-predicted **cavitation** patterns and stroboscopic photographs of cavitation. The results serve as a proof of concept of the combined CFD-acoustic analogy approach to modelling radiated noise from cavitating propellers and simultaneously highlight deficiencies of the present methodology.

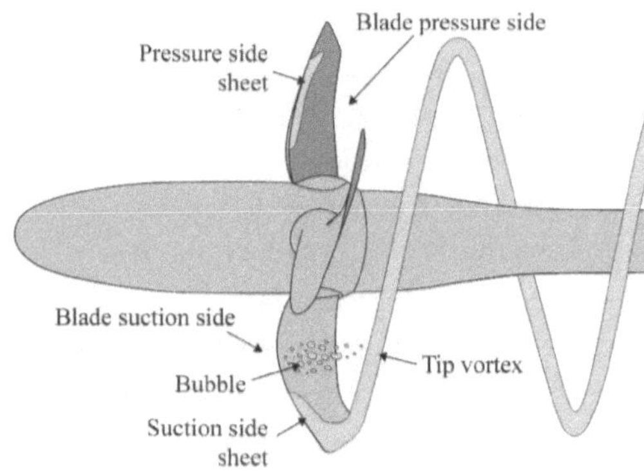

Figure 2.1: *Schematic of the four types of propeller-induced cavitation considered. Flow is from left to right.*

The present analysis shares many features of its methodology with Sezen et al. [44], including the same experimental facility and computational modelling strategy. Howver, this study differes in significant ways. Fist the propellers studied are quite different in design; Sezen et al. [44] studied a high-solidity fixed-pitch propeller, whereas the present work studies a low-solidity variable pitch model and includes the effects of pitch in the analysis. Second, the present work examines pressure side cavitation, which was not studied in Sezen et al. [44]. Finally, the frequency range of interest in the present work is higher, including frequencies up to 10^5 Hz and excluding low frequencies.

2.4.2 Experimental system and techniques

2.4.2.1 Flow facility

Model-scale propeller tests were conducted in a closed-circuit cavitation tunnel at the University of Genoa, illustrated schematically in Figure 2.2. Uniform inflow with a maximum velocity of 8.5 m/s was generated at the entrance of a 2.2 m-long test section that had a cross-section of 0.57 m x 0.57 m. The concentration of dissolved oxygen was maintained at 4.5 ppm, which was found to provide sufficient seeding for cavitation while simultaneously minimizing noise absorption by free bubbles.

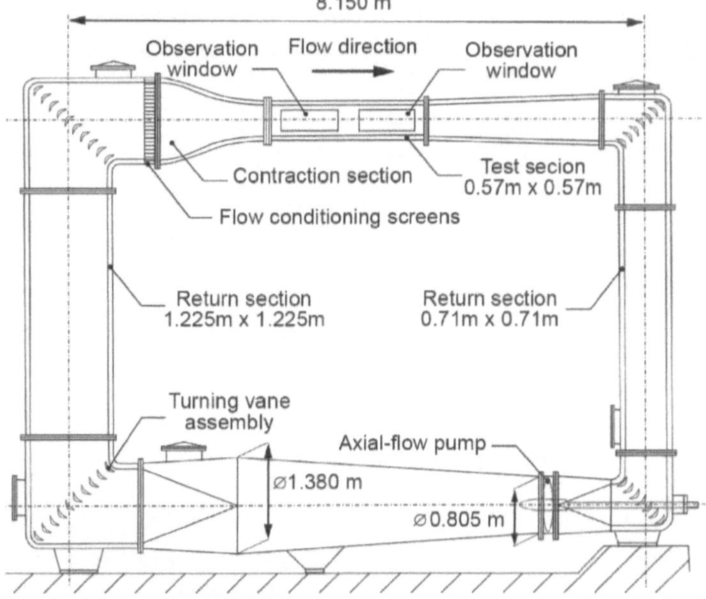

Figure 2.2: Schematic of the cavitation tunnel.

A scale model propeller was positioned near the inlet of the test section with its supporting pod located downstream, as shown in Figure 2.3, to ensure uniform inflow conditions.

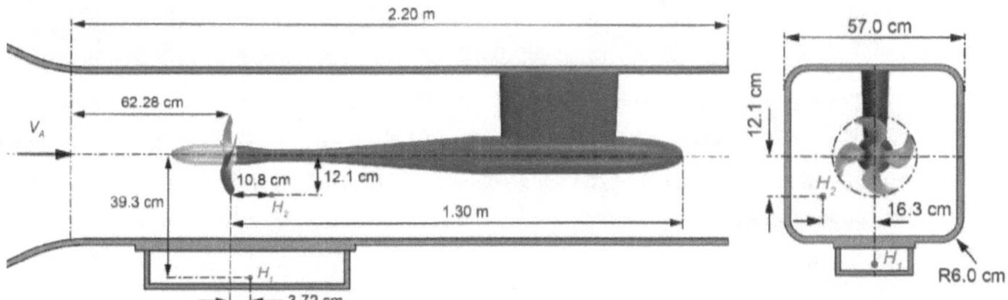

Figure 2.3: Close-up schematic of the test section. Left: streamwise view; right: transverse view. Points H_1 and H_2 represent the locations of the hydrophones.

2.4.2.2 Propeller model

The propeller used in the present experiments was a scale model of a CPP representative
of a mid-size tanker vessel. The parameters of the propeller are presented in Table 2.1.
Two pitch configurations were studied, referred to hereafter as the design pitch and the
reduced pitch.

Table 2.1: Propeller parameters

Number of blades	4
Model diameter	0.24 m
Design pitch $(P/D)_{0.7R}$	0.87
Reduced pitch $(P/D)_{0.7R}$	0.521

Propeller rotation speed, thrust, and torque were measured with a Kempf & Remmers H39
dynamometer contained within the pod and corrected for tunnel effects using the
corrections of Wood and Harris [51].

2.4.2.3 Acoustic measurements

Acoustic measurements were performed using a pair of miniaturized active hydrophones,
a Bruel & Kjaer type 8103 (H_1) and a Reson TC4013 (H_2), positioned as shown in Figure
2.3. Hydrophone H_1 was submerged in water and separated from the test section by a
plexiglass window, while hydrophone H_2 was located inside the test section. The time-
domain sound pressure signals consisted of 2^{21} (approximately 2.1 billion) samples at 200
kHz.

Post-processing of acoustic measurements was conducted according to the International
Towing Tank Conference (ITTC) guidelines for model scale noise measurements [52].
Acoustic power spectral densities $G(f)$ were obtained via Welch's method of averaging for
modified spectrograms [53]. Sound levels in dB (1 μPa^2/Hz reference) were used, defined
according to:

$$L_P(f) = 10 \log_{10}\left(\frac{G(f)}{P_{\text{ref}}^2}\right).$$

(2.1)

Background noise was measured replacing the propeller with a dummy hub and running
the facility at the same operational conditions of propeller tests. The background noise
was used to correct the noise levels computing the net sound levels. The background

correction depended on the ratio of net noise level to background noise level, where the latter was measured independently. For signal-to-noise ratios greater than 3 dB, the background noise level was subtracted from total noise level according to:

$$L_{P,\text{net}} = 10 \log_{10} \left(10^{(L_{P,\text{total}}/10)} - 10^{(L_{P,\text{background}}/10)} \right). \tag{2.2}$$

Any portions of measured spectra for which the signal-to-noise ratio was less than 3 dB were excluded from the analysis.

Final noise signals were presented in terms of the non-dimensional pressure coefficient K_P, in which the pressure was normalized by the diameter and the propeller revolution rate, according to:

$$L_{KP}(f) = 20 \log_{10} \left(\frac{K_P(f)}{10^{-6}} \right), \quad K_P = \frac{p}{\rho n^2 D^2} \tag{2.3}$$

where p is the local pressure, ρ is the liquid density, n is the revolution rate of the propeller (in revolutions per second), and D is the propeller diameter.

The present experimental results were not scaled to full-scale, since the goal of the study was only fundamental insight and numerical replication of the measured acoustic signal. In principal, scaling of acoustic data from model-scale propeller experiments is possible. The ITTC guidelines for model-scale cavitation noise measurements present scaling procedures for cavitation-induced noise; however, the accuracy of the scaling procedure is difficult to determine due to the impossibility of achieving full dynamic similarity, difficulty in isolating the cavitation noise source at full-scale, and measurement uncertainties at both model- and full-scale [52].

2.4.2.4 Stroboscopic flow imaging

The flow field in the vicinity of the propeller was illuminated by two stroboscopic lights (900 Movistrob). Images of the cavitation patterns were recorded using three Vision Tech Marlin F145B2 Firewire cameras with a resolution of 1392 by 1040 pixels and frame rate up to 10 fps. We obtained three concurrent views of the propeller: the pressure side and the suction side of a single blade, as well as a view of the entire propeller.

2.4.2.5 Operating conditions

Ten propeller loading conditions, five for each pitch setting, were tested by varying the advance ratio and the cavitation index to induce each of the cavitation regimes shown in

Figure 2.1. Flow conditions were characterized according to the advance ratio $J = V_A/nD$ and the nominal cavitation index σ_N:

$$\sigma_N = \frac{p_{ref} - p_v}{\frac{1}{2}\rho n^2 D^2}, \tag{2.4}$$

where V_A is the advance velocity, p_{ref} the hydrostatic pressure at the propeller shaft depth after depressurization of the tunnel and p_v was the vapour pressure of water at the test conditions. Experimental conditions were scaled to ensure cavitation index similarity with the full-scale. Specific conditions were achieved by varying the inflow velocity and the level of depressurization of the cavitation tunnel. Several noise measurements were carried out for each operating condition, after stopping and restarting the cavitation tunnel, to ensure repeatability of the results. Further, a sensitivity study was performed by inducing small percentage variations to controlled parameters to check measurement uncertainty according to the ITTC guideline [52].

The experimental conditions and the corresponding cavitation regimes are summarized in Table 2.2. Pitch setup is specified by the diametral pitch at 70% of the propeller radius $(P/D)_{0.7R}$. Measured loading on the propeller loading is presented in terms of thrust coefficient $K_T = F_T/\rho n^2 D^4$ and torque coefficient $K_T = Q/\rho n^2 D^5$, where F_T, and Q are the propeller's thrust and torque respectively. Conditions C1, C2, and C3 were representative of the propeller operating at its design pitch and advance ratio under different ambient pressure conditions. Conditions C4 and C5 represented off-design loading conditions. Condition C4 represented a low propulsive loading condition, and it was achieved by increasing the advance ratio. Condition C5 represented the opposite case, where high propulsive loading and slip resulted from a low advance ratio. Conditions C1b through C5b were the reduced pitch conditions, otherwise analogous to their design pitch counterparts.

Table 2.2: Operating conditions and the corresponding cavitation regimes.

Condition	$(P/D)_{0.7R}$	J	σ_N	n [rps]	K_T	$10K_Q$	Cavitation type
C1	0.87	0.516	2.9	25	0.205	0.293	Tip vortex
C2	0.87	0.516	2.3	25	0.205	0.293	Tip vortex
C3	0.87	0.516	1.4	25	0.205	0.293	Tip vortex, suction side sheet, bubble
C4	0.87	0.769	2.3	25	0.09	0.172	Pressure side leading edge
C5	0.87	0.345	2.3	25	0.27	0.350	Tip vortex, suction side sheet
C1b	0.521	0.404	2.6	30	0.095	0.125	Pressure side leading edge
C2b	0.521	0.404	2.3	30	0.095	0.125	Pressure side leading edge
C3b	0.521	0.404	1.4	30	0.095	0.125	Pressure side leading edge
C4b	0.521	0.500	2.3	30	0.05	0.095	Pressure side leading edge
C5b	0.521	0.345	2.3	30	0.12	0.140	None

2.4.3 Numerical simulations

2.4.3.1 Hydrodynamic model

Flow is modelled in STAR-CCM+ (12.04.011-R8) using the incompressible URANS equations. The realizable k-ε two-equation turbulence model was used in the present simulations. The realizable k-ε model modifies the standard k-ε model in two ways: first, the Reynolds stresses are replaced in the dissipation rate equation by a "source" term; and second, a different eddy viscosity equation is used which ensures realizability and accounts for the effects of mean flow rotation. The result is a model that has improved performance for flows with high mean shear rates or large separation regions [54].

Turbulence modelling in the present simulations is discussed further in Appendix A.

A mixture (volume-of-fluid) approach is taken for modelling the two-phase flow resulting from cavitation, where the fluid is treated as a homogeneous mixture with bulk properties determined at the individual cell level according to the vapour volume fraction α. Each of the density, viscosity, and velocity components was calculated as a weighted average of the vapour and liquid phase properties. For any given property B, the mixture property is computed according to $B_{\mathrm{mix}} = \alpha B_{\mathrm{vap}} + (1 - \alpha)B_{\mathrm{liq}}$. In this representation, the URANS equations take on a mixture form (Eqn. (2.4)), dependent upon the interphasic mass transfer rate $\dot{m}$ (Eqn. (2.6)). In Einstein notation, adopting u as a velocity component and x as a spatial coordinate, and using overbars for time-averaged quantities and primed notation for fluctuating components the mixture form of the URANS equations is:

$$\frac{\partial\left(\rho_{\mathrm{mix}}\bar{u}_{i,\mathrm{mix}}\right)}{\partial t} + \frac{\partial}{\partial x_j}\left(\rho_{\mathrm{mix}}\bar{u}_{i,\mathrm{mix}}\bar{u}_{j,\mathrm{mix}} + \rho_{\mathrm{mix}}\overline{u'_{i,\mathrm{mix}}u'_{j,\mathrm{mix}}}\right)$$
$$= -\frac{\partial\bar{p}}{\partial x_i} + \frac{\partial}{\partial x_j}\left[\mu_{\mathrm{mix}}\left(\frac{\partial\bar{u}_{i,\mathrm{mix}}}{\partial x_j} + \frac{\partial\bar{u}_{j,\mathrm{mix}}}{\partial x_i}\right)\right], \tag{2.5}$$

where

$$\frac{\partial\bar{u}_i}{\partial x_i} = \dot{m}\left(\frac{1}{\rho_{\mathrm{mix}}} - \frac{1}{\rho_{\mathrm{vap}}}\right) \tag{2.6}$$

and

$$\dot{m} = \frac{\rho_{\mathrm{vap}}\rho_{\mathrm{liq}}}{\rho_{\mathrm{mix}}}\frac{d\alpha}{dt}. \tag{2.7}$$

The interphasic mass transfer was modelled with the Schnerr and Sauer cavitation model, in which the vapour phase is assumed to consist of a collection of N spherical bubbles of radius R that individually behave according to the Rayleigh equation [55]:

$$\frac{dR}{dt} = \sqrt{\frac{2}{3}\frac{|p_B - p_\infty|}{\rho_{\mathrm{liq}}}}. \tag{2.8}$$

In the context of the model, the internal bubble pressure p_B is assumed be uniform and equal to the vapour pressure of water, while the ambient pressure p_∞ is taken to be the ambient cell pressure. The spherical bubble assumption allows the vapour fraction to be expressed in terms of the radius, leading to a differential equation relating the rates of change of the vapour fraction and the radius:

$$\alpha = \frac{(4/3)N\pi R^3}{\mathcal{V}_{\mathrm{liq}} + (4/3)N\pi R^3}, \tag{2.9}$$

$$\frac{d\alpha}{dt} = \alpha(1 - \alpha)\frac{3}{R}\frac{dR}{dt}, \tag{2.10}$$

where $\mathcal{V}_{\text{liq}}$ is the total liquid volume in the cell. Combining Eqns. (2.7), (2.8), and (2.10) yields a useful form of the interphasic mass transfer equation:

$$\dot{m} = 3\frac{\rho_{\text{vap}}\rho_{\text{liq}}}{\rho_{\text{mix}}}\frac{\alpha(1-\alpha)}{R}\sqrt{\frac{2}{3}\frac{|p_B - p_\infty|}{\rho_{\text{liq}}}}. \tag{2.11}$$

It is noteworthy that the Schnerr and Sauer cavitation model relies on a significant simplification of the physics of bubble oscillation, which neglects nucleation and collapse entirely. The use of the model in CFD is motivated primarily by its simplicity and closed form. Realistic cavitation bubbles are not entirely filled with saturated water vapour, but instead contain a mixture of vapour and gas. Polytropic expansion and compression of gas in the bubbles leads to thermal damping, which is ignored in the model. Further, the Rayleigh relation is a simplification of the Rayleigh-Plesset equation that describes the dynamics of spherical bubbles in an infinite body of incompressible fluid [32]. Assuming a bubble is filled entirely with saturated vapour, the Rayleigh-Plesset equation takes the form:

$$\rho\left[R(t)\frac{d^2R(t)}{dt^2} + \frac{3}{2}\left(\frac{dR(t)}{dt}\right)^2\right] = p_v - p_\infty(t) - \frac{2\gamma_s}{R(t)} - 4\rho v\frac{1}{R(t)}\frac{dR(t)}{dt}, \tag{2.12}$$

where ρ is the density of the liquid, $R(t)$ is the bubble radius, p_v is the vapour pressure of the liquid, $p_\infty(t)$ is the ambient pressure, γ_s is the surface tension of the bubble envelope, v is the kinematic viscosity of the liquid, and t is time. The Rayleigh-Plesset equation, in turn, neglects the effects of acoustic radiation into the surrounding fluid, which serves as an additional source of damping. Moreover, the Rayleigh relation assumes a large pressure difference $p_B - p_\infty$ and the dominant role of inertia effects in bubble growth, eliminating the surface tension and the viscous damping terms [56]. Therefore, all sources of damping in cavitation bubble oscillation are neglected in the Schnerr and Sauer model.

2.4.3.2 Hydroacoustic model

Hydroacoustic behaviour was modelled with the porous surface solution to the Ffowcs Williams-Hawkings (FWH) acoustic analogy equation first proposed by di Francescantonio [57], who combined the work of Farassat on Kirchhoff formulations [58] with the FWH. This technique was applied to CFD simulations of cavitating propellers by Ianiello and de Bernardis [40] and is outlined in the following.

The acoustic analogy of Ffowcs Williams and Hawkings is a rearrangement of the conservation laws for mass and momentum into an inhomogeneous wave equation for density perturbations with quadrupole, dipole, and monopole source terms [59]:

$$\left(\frac{\partial^2}{\partial t^2} - c^2 \frac{\partial^2}{\partial x_i^2}\right)(\overline{\rho - \rho_0}) = \frac{\partial^2 \overline{T_{ij}}}{\partial x_i \partial x_j} - \frac{\partial}{\partial x_i}\left(p_{ij}\delta(f)\frac{\partial f}{\partial x_j}\right) + \frac{\partial}{\partial t}\left(\rho_0 v_i \delta(f)\frac{\partial f}{\partial x_i}\right). \qquad (2.13)$$

Closed mathematical surfaces coincident to real moving surfaces within the flow that constitute the physical sources of sound, allowing the construction of the three source terms in Eqn. (2.13). The first of the three terms on the right-hand side of Eqn. (2.13) describe quadrupole sources that lie outside any moving surfaces. Here, the term $\overline{T_{ij}}$ is equal to the Lighthill stress tensor $T_{ij} = pu_i u_j + p_{ij} - c^2(\rho - \rho_0)\delta_{ij}$ outside of these surfaces, and it is equal to zero within them (δ_{ij} is the Kronecker delta). The function $f(\mathbf{x}, t)$ is defined such that $f = 0$ and $\Delta f = 1$ on the surfaces, and thus the second and the third source terms in Eqn. (2.13) are zero-valued, except on the surfaces. These two terms represent the dipole source distribution and the monopole source contribution due to the surface movement, respectively.

The classical FWH equation relies on coincidence of the mathematical surfaces with real surfaces. By combining Eqn. (2.13) with Farassat's formulation, di Francescantonio gives a solution that allows for an arbitrarily defined porous surface S_p that is restricted only in that it must contain all sources of noise and must be located far from those sources [57]. To achieve this form of the solution, alternate velocity and stress terms are defined as follows:

$$U_i = \left(1 - \frac{\rho}{\rho_0}\right)v_i + \frac{\rho}{\rho_0}u_i, \quad L_i = p_{ij}\hat{n}_j + \tilde{\rho}u_i(u_n - v_n), \qquad (2.14)$$

where v and u are the surface and fluid velocities, respectively, p_{ij} is the compressive pressure stress tensor, $\hat{n}_j$ is the j component of the porous surface-normal unit vector, and $\tilde{\rho}$ is the density disturbance due to acoustic perturbation. The n subscript on the velocities here denotes the surface-normal component. Changing the Eqn. (2.14) into an integral form by application of the free-space Green's function gives the solution form useful for numerical computation of acoustic pressure in terms of the source-to-observer distance r, the projection of the local Mach number along the source-to-observer vector M_r, as well the newly-defined velocity and stress terms. This form was referred to as the Kirchoff FWH (KFWH) equation:

$$4\pi p(\mathbf{x}, t) = \frac{\partial}{\partial t}\int_{S_\mathrm{p}}\left[\frac{\rho_0 U_n}{r|1 - M_r|}\right]dS_\mathrm{p} + \frac{1}{c_0}\frac{\partial}{\partial t}\int_{S_\mathrm{p}}\left[\frac{L_r}{r|1 - M_r|}\right]dS_\mathrm{p} + \int_{S_\mathrm{p}}\left[\frac{L_r}{r^2|1 - M_r|}\right]dS_\mathrm{p} . \qquad (2.15)$$

Compared to the original FWH analogy, the only additional assumptions that have been made in deriving Eqn. (2.15) are the restrictions on the surface, however the physical significance of the individual terms describing quadrupole, dipole, and monopole sources is eliminated. The advantage of this formulation as a hydroacoustic model in the context of CFD is that acoustic analysis becomes pure post-processing of the hydrodynamic solution on the porous surface.

2.4.3.3 Computational domain and boundary conditions

A pair of overset meshes comprised the computational domain. One stationary mesh, shown in Figure 2.4, was a geometric replica of the test section of the cavitation tunnel and the propeller pod. The second mesh was used as a rotating domain and contained the propeller rotor; it is shown as a transparent cylinder in Figure 2.4. Physical boundary conditions of uniform inflow and ambient pressure were selected to match the corresponding experimental conditions, with boundary conditions on solid surfaces treated using k-ω wall functions. The outer surface of the rotating domain was also used as the porous integration surface for the KFWH acoustic solution. Total pressure and the KFWH solution were sampled at the location of the physical hydrophone H2, shown in Figure 2.3, which was located within the porous surface in the computational domain.

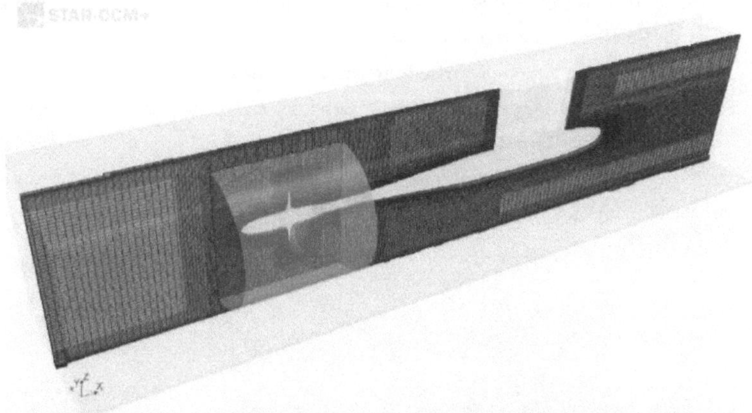

Figure 2.4: Isometric view of the computational domain and mesh. The cylinder surrounding the propeller rotor represents both the boundary of the rotating domain and the integration domain for the FWH equation solution.

The numerical simulations were conducted in three stages. First, a hydrodynamic solution was allowed to evolve with the cavitation model turned off until a stable condition was achieved. The cavitation model was then activated and the simulation allowed to reach a new stable condition. Final data, including the KFWH acoustic pressure solution, were obtained from one-quarter revolution of the propeller for each loading condition, thereby capturing the hydrodynamic behaviour throughout one full rotation of the propeller.

2.5   Results and Discussion

2.5.1 Propeller loading

The URANS solutions were validated by comparing the calculated thrust and torque coefficient values with the dynamometer measurements from the experimental campaign. As shown in Table 2.3, most simulations approximated torque and thrust to within 10% of the experimental measurements. The experimental conditions and the corresponding cavitation regimes are defined in Table 2.2. The worst match was observed between experiments and simulations for the two high-advance-ratio conditions, C4 and C4b. In these conditions the local angle of attack at each radial position along the propeller blades was the smallest of all tested conditions, resulting in low pressure loading and increasing the relative contribution of vorticity to the overall propeller load. The mismatch in loading values was likely the result of a loss of vortex feature fidelity in the propeller wake in the URANS solutions due to numerical damping, as well as the presence of a laminar boundary layer in the scale model experiments, as discussed by Bulten [60], that was not modelled in the simulations.

Table 2.3: Comparison of measured and calculated thrust and torque coefficients

Condition	Experimentally Measured		URANS			
	K_t	K_q	K_t	Error	K_q	Error
C1	0.205	0.0293	0.195	5%	0.0288	2%
C2	0.205	0.0293	0.195	5%	0.0288	2%
C3	0.205	0.0293	0.195	5%	0.0288	2%
C4	0.090	0.0172	0.076	15%	0.0158	8%
C5	0.270	0.0350	0.270	0%	0.0356	2%
C1b	0.095	0.0125	0.091	4%	0.0123	2%
C2b	0.095	0.0125	0.091	4%	0.0123	2%
C3b	0.095	0.0125	0.091	4%	0.0123	2%
C4b	0.050	0.0095	0.045	10%	0.0090	5%
C5b	0.120	0.0140	0.116	3%	0.0139	1%

2.5.2 Cavitation patterns during design pitch operation

Four distinct cavitation regimes were investigated in the cavitation tunnel experiments at the design pitch setting of the propeller. Stroboscopic images of these cavitation regimes are presented in Figure 2.5 alongside their numerically predicted counterparts. Only solutions that used the realizable k-ε turbulence model are presented in the current section; the turbulence modelling effects are discussed separately. Three views of the flow field were captured during the experiments using stroboscopic flow visualization: views of the suction and the pressure sides of a single blade and a view of the entire rotor. No experimental condition exhibited cavitation on both sides of a propeller blade, and the images in Figure 2.5 show all cavitation observed at each condition. Each image shows an instantaneous shape of the cavities. For stable cavities, the images are representative of any typical moment in time. However, bubble-type cavities formed and collapsed sporadically, and therefore cannot be fully represented with instantaneous images.

Under the design pitch and loading (represented by the advance ratio) conditions at high and moderate cavitation indices (conditions C1 and C2), stable tip vortex cavitation developed in near isolation, i.e. in the absence of other types of cavitation. Attached sheet cavities covered insignificant surface area of the blades and were localized to the vicinity of the blade tips. Decreasing the cavitation number (condition C3) resulted in simultaneous formation of both sheet and bubble cavities, in addition to tip vortex cavities. The decrease in the cavitation number resulted in a thicker tip vortex cavity. It was visually observed that the tip vortex cavities extended for several propeller diameters downstream of the propeller, beyond the field of view of the stroboscopic imaging system.

When the advance ratio was increased (condition C4), the tip vortex cavitation was suppressed, and a sheet cavity developed along the leading edge on the pressure side of the blade. When the advance ratio was decreased relative to the design condition (condition C5), the tip vortex cavity expanded in the radial direction, and it was accompanied by a large sheet cavity on the suction side of the blade. Condition C5 produced the largest sheet cavitation among all tested loading conditions. It should be noted that in Figure 2.5 it appears that two separate vortex cavities developed from the blade tip; however, alternate views of the flow field show that the second cavity formed at a different blade.

Visual representation of cavities predicted by the volume of fluid method in CFD is somewhat arbitrary since the vapour cavities are represented by a distribution of small bubbles, with no defined envelope. On the other hand, in the experiments it was not feasible to determine the threshold of the bubble density at which the cavitation became visible in a stroboscopic photograph. Therefore, the visualization threshold was adjusted for the numerical results. It was found that a threshold of approximately 1% vapour by volume resulted in reasonable reproduction of cavity shapes. At this threshold, the thickness of tip vortex cavities near their inception was qualitatively similar to the experimental observations as shown in Figure 2.6.

The extent of tip vortex cavities was under-predicted by the URANS solutions in conditions C1, C2, C3, and C5, as shown in Figure 2.6 for condition C2. This effect is a known limitation of URANS formulations, resulting from numerical damping that can be resolved by performing a DES [40]. However, DES solutions are significantly more computationally expensive than URANS, and extension of this study to DES was therefore left to future work.

The URANS solutions also failed to predict the existence of bubble cavitation under loading condition C3. The solution instead exhibited a large sheet cavity along the surface of the blades, as shown in Figure 2.5. Since the Schnerr and Sauer model uses a simplified version of the Reighleigh-Plesset equation, which itself assumes the cavitation in the form of a spherical bubble in an infinite medium, the inability to predict the violent expansion and collapse of bubbles was expected. It was unclear if the surface area of the blade on which a sheet cavity was predicted by the URANS corresponded to the region on which bubble cavities developed in the experimental campaign.

The position and spatial extent of sheet cavitation was accurately predicted by the URANS solutions. The pressure side cavity produced under condition C4 was very similar in extent between the experiments and the simulations. However, the cavity remained attached to the surface of the blade in the URANS solution rather than separating toward the tip of the blade, as it was seen in the experiments. Moreover, a thin cavity extending along the leading edge of the blade that was not seen in the experiments appeared in the simulations of conditions with sheet cavitation, which can be clearly seen in the case of condition C5 in Figure 2.5. In the model scale experiments a laminar boundary layer is likely to postpone the occurrence of sheet cavitation at inner radii, but the laminar boundary layer is not captured in the present Reynolds-Averaged Navier-Stokes (RANS) scheme.

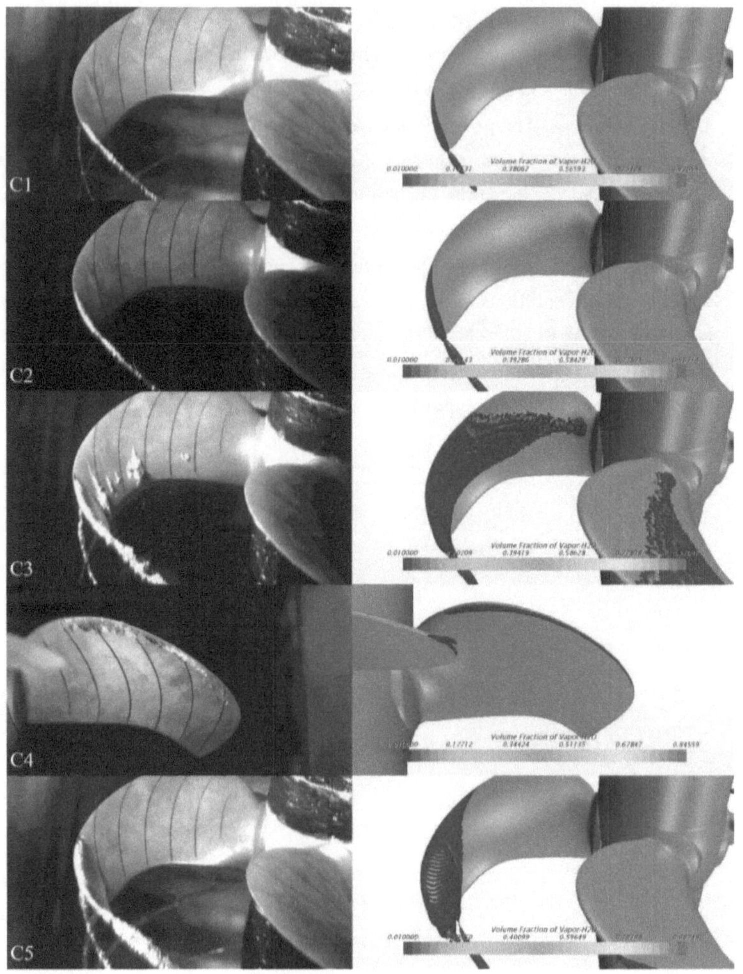

Figure 2.5: *Cavitation patterns corresponding to the design pitch of the propeller blades: instantaneous stroboscopic images (left) and URANS solution (right). For all conditions, except C4, the suction side of the blade is shown. The pressure side of the blade is shown for condition C4.*

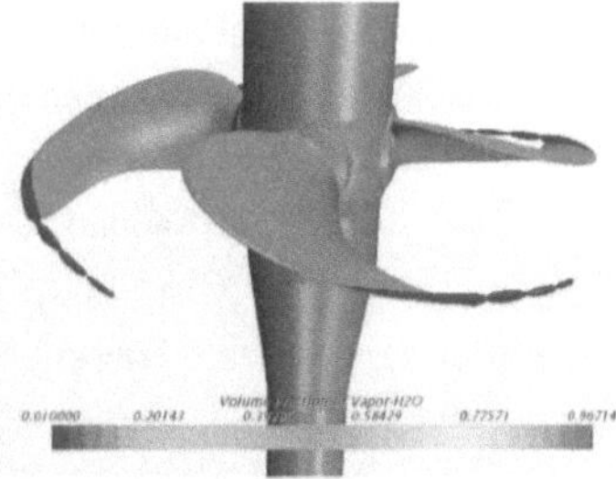

Figure 2.6: Cavitation pattern in the near-wake of the propeller corresponding to the design pitch of the propeller blades at the condition C2: Instantaneous stroboscopic photograph (left) and URANS solution (right).

2.5.3 Cavitation patterns during reduced pitch operation

Under the reduced pitch experimental conditions, only pressure side sheet cavitation on the leading edge was observed, as shown in Figure 2.7. The expected trend of increasing cavitation with decreasing the cavitation number was observed during the three conditions (C1b, C2b and C3b) with the same thrust and torque loading on the propeller as the design pitch conditions C1, C2 and C3. Among the three conditions at an identical cavitation number (C2b, C4b and C5b), a trend of increasing cavitation with decreasing thrust and torque load was observed. Condition C5b, which had the highest propeller loads of all reduced pitch cases, exhibited complete suppression of cavitation.

Similar to the design pitch condition C4, the URANS simulations performed well in terms of reproduction of the pressure side sheet cavitation patterns. Conditions C1b, C2b, and C4b showed attached cavities along the leading edge of the blades that matched the experimentally obtained patterns. Moreover, the simulations reproduced the same cavitation pattern at the tip of the blade that was observed in the simulation of condition C4. At the low cavitation number (condition C3b), the URANS solution developed a large sheet cavity on the pressure side of the blade that was not observed experimentally. Two factors likely explain the over-prediction of cavitation on the suction side of the blades in this case. First, the simulations may under-predict the pressure on the suction side of the blades due to the RANS representation of the boundary layer, leading to the inception of bubble cavitation at a lower cavitation number. Second, the numerical model does not capture the complete bubble dynamics, causing bubble cavitation to be represented by a large sheet cavity. Cavitation was also erroneously predicted by the simulation of condition

C5, perhaps due to uncaptured boundary layer effects that supressed cavitation in the model-scale experiments, although the predicted volume of cavitation at that condition was the lowest among the reduced-pitch cases.

The change of cavitation regime between cases C1, C2, C3 and C5 and their reduced-pitch counterparts C1b, C2b, C3b and C5b confirms, as expected, that pressure side cavitation is generally associated with both high advance ratios and reduced pitch operation, even taking into account the scaling of the advance ratio with pitch. Since the spanwise twist distribution of propeller blades is optimized for the design pitch operation, it is likely that suboptimal twist distribution at reduced pitches serves to promote vortex development at the pressure side of the leading edge, resulting in pressure side cavitation. Increasing the trust and torque load by decreasing the advance ratio supresses this phenomenon, as it was observed at the condition C5b.

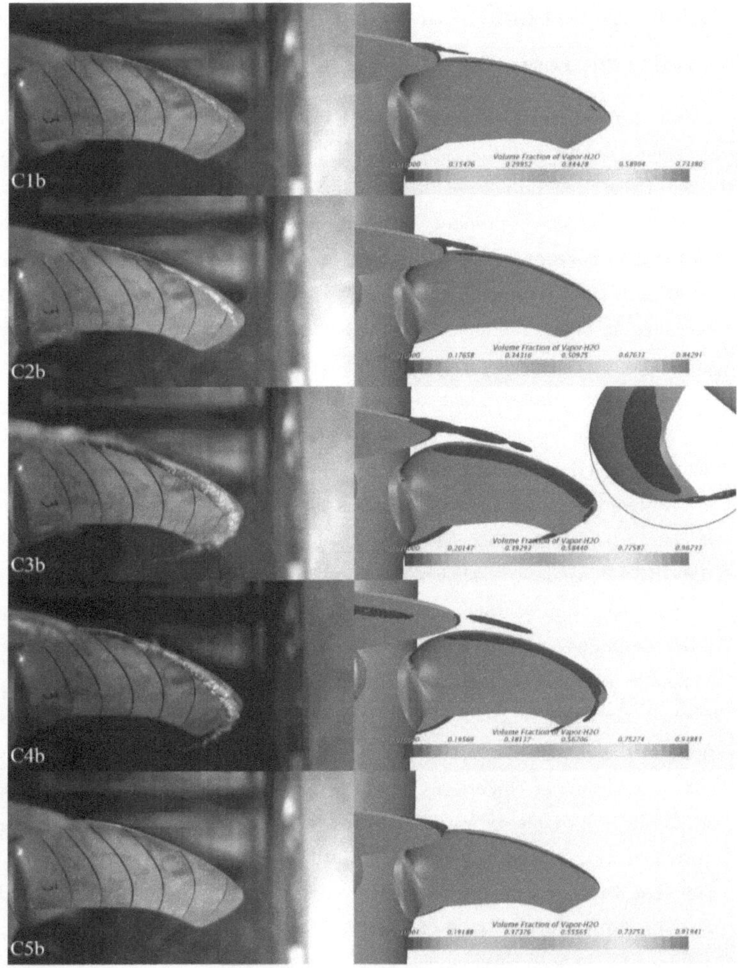

Figure 2.7: Cavitation patterns corresponding to the reduced pitch of the propeller blades: instantaneous stroboscopic images (left) and URANS solution (right). For all conditions, the pressure side of the blade is shown. The pressure side of the blade is shown for condition C4. The cut-out image shows the suction side of the blade for condition C3b.

2.5.4 Experimental acoustic signatures of the observed cavitation regimes

Power spectral density measurements of a non-cavitating condition corresponding to the propeller loading conditions C1, C2 and C3 was compared with the background tunnel noise. Except for the noise at the blade passing frequency of 100 Hz, the non-cavitating spectrum contained primarily background noise. Therefore, the application of a background noise correction effectively served to isolate noise from cavitation sources at frequencies other than the blade passing frequency. All spectra presented herein have been corrected for background noise as described in Section 2.4.3.1. It should be noted that gaps in the spectra, which occur primarily at low frequencies, are the result of excluding data at frequencies with low signal-to-noise ratios.

The unique acoustic signatures of each of the four cavitation regime combinations observed experimentally can be identified from the power spectral density measurements of the five design pitch test conditions presented in Figure 2.8 in terms of the non-dimensional pressure coefficient L_{KP}, which is defined by Eqn. (2.3). A line representing the 4 dB/octave decay is shown in Figure 2.8 as a reference.

Each spectrum contains a peak at the blade passing frequency, which for the design pitch conditions was equal to 100 Hz. The two conditions under which tip vortex cavitation formed in isolation, C1 and C2, emitted sound characterised by a broad hump, the centre frequency of which ranged from 400 Hz to 800 Hz, and logarithmically decaying broadband noise at frequencies above 1 kHz. The width and the maximum amplitude of the hump were greater under condition C2, which exhibited thicker vortex cavitation structures, while the peak frequency was reduced relative to condition C1. This component of the tip vortex cavitation noise is associated with the oscillation of the cavitating vortex core [50], [34]. The decay of acoustic power observed in the higher frequency broadband component of the noise was in the range of 4 to 4.5 dB/octave, which is within the range commonly assumed by semiempirical models [61].

Condition C5, under which a combination of suction side sheet cavitation and tip vortex cavitation developed, resulted in noise emission with a similar signature to conditions C1 and C2 at a higher overall noise level and with a hump occurring at lower frequencies. However, the relative magnitude of the hump in comparison to the broadband noise level was less than that observed in the cases with minimal sheet cavitation.

The low frequency noise signature of the condition C3, under which bubble cavitation formed, was similar to that of conditions C1, C2, and C5. At high frequencies, sound levels were higher than those produced under any other studied condition due to the presence of bubble cavitation.

No hump was observed in the spectrum corresponding to the condition C4, under which tip vortex cavitation was suppressed and replaced by pressure side sheet cavitation. Instead, the spectrum was characterised by a peak amplitude occurring at a frequency above 1 kHz and a logarithmic decay trend that extended to approximately 20 kHz, similar to the decay observed at other experimental conditions. Unlike the spectra recorded at the other design pitch conditions, the slope of the decay trend increased markedly above 20 kHz during the condition C4.

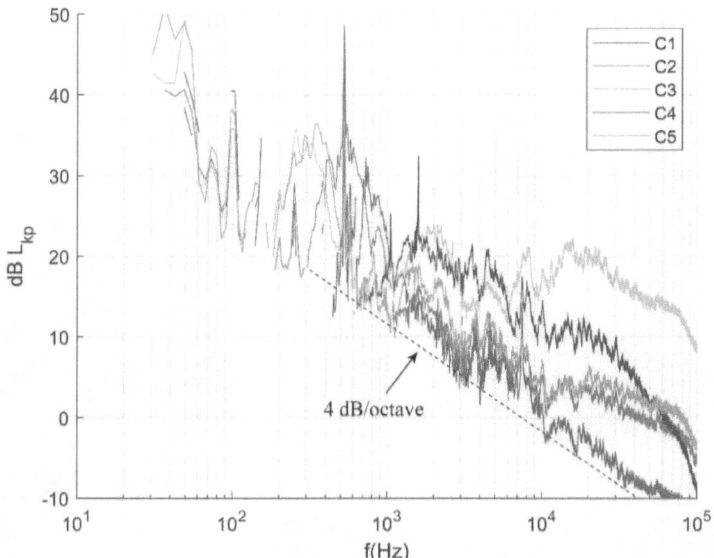

Figure 2.8: Net power spectral density corresponding to the design pitch of the propeller blades measured by the hydrophone H2.

Acoustic spectra for the propeller operating under the reduced pitch conditions are provided in Figure 2.9. All acoustic spectra exhibited qualitatively similar patterns to those of the condition C4, particularly in the frequency range above 1000 Hz. This result is

expected, considering that pressure side cavitation was predominant under conditions C1b
through C4b. The acoustic amplitude differed between the cases, roughly corresponding to
the cavity size observed in the stroboscopic images. Among the reduced pitch conditions,
only conditions C1b and C2b exhibited the abrupt change in the slope of the decay at the
high frequencies that was characteristic of condition C4. Condition C4b exhibited a unique
peak at approximately 500 Hz that was similar to the hump observed at the conditions
with tip vortex cavitation, although the source of this tonal noise component in condition
C4b was unclear. This tonal noise component is likely associated with a pulsation
phenomenon similar to that previously described for the tip vortex, but now occurring on
the cavity of the pressure side vortex. This phenomenon may occur also on pressure side
vortices when their extent allows the formation of a stable cavity.

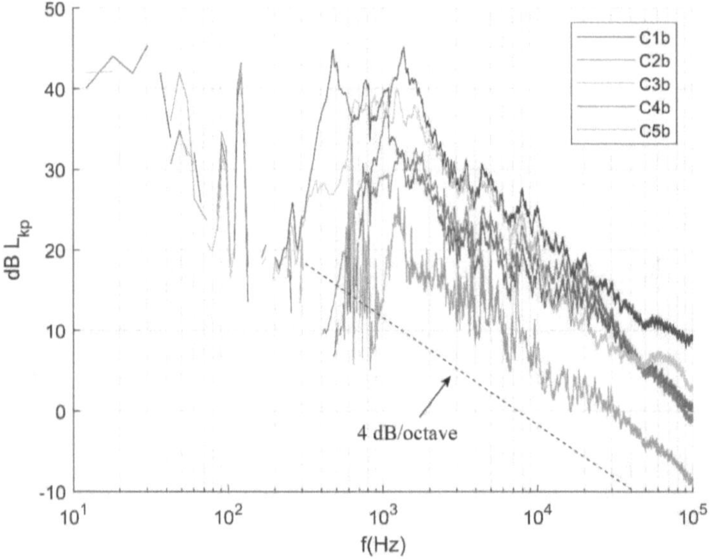

*Figure 2.9: Net power spectral density corresponding to the reduced pitch of the propeller blades
measured by the hydrophone H2.*

While the decay trend of the high-frequency broadband component of the power spectra
was preserved between the majority design and reduced pitch cases, most of the reduced
pitch conditions showed increases in the sound pressure levels of up to 20 dB relative to

the corresponding design pitch conditions. There were two exceptions to this trend: high sound levels associated with bubble cavitation resulted in increased high-frequency sound levels under condition C3 compared to condition C3b, and the suppression of all types of cavitation at condition C5b resulted in the lowest overall sound level. In all cases where the design pitch condition resulted in tip vortex and/or suction-side sheet cavitation while its reduced pitch counterpart produced pressure-side sheet cavitation, the reduced pitch condition was the louder of the two. This result is counterintuitive, considering that the reduced pitch tests were performed at lower freestream velocities, thrust coefficients, and torque coefficients. However, pressure side cavitation, which is less likely to occur during design operating conditions, is expected to result in significantly higher levels of noise than comparable suction side cavitation [61].

This result highlights the need for intelligent operating strategies to reduce anthropogenic marine noise. Rudimentary control strategies that assume linear ship velocity-to-noise relations are inadequate for vessels with controllable-pitch propellers. For example, in the case of a decelerating ship with a controllable pitch propeller, keeping the engine speed fixed while adjusting the propeller pitch will often result in excessive noise levels [55]. A more appropriate speed-reduction strategy would be to maintain the effective angle of attack of the propeller blades close to the design values to reduce cavitation, especially on the pressure side of the blades. Further, the reduction of blade pitch results in suboptimal span-wise twist distributions, which makes a simple scaling of the pitch with the advance ratio insufficient for prediction of the radiated noise. The present results indicate that a bias toward low advance ratios and increased propeller loading during low-speed operation should result in lower acoustic emission through suppression of pressure-side cavities.

2.5.5 Numerically predicted power spectra of the radiated acoustic noise

Power spectra were computed using two approaches. The first approach involved a direct calculation of the pressure values from the grid cell corresponding to the location of the hydrophone H2 using the URANS equations. The second approach involved application of a KFWH acoustic analogy to calculate the pressure values at the same location. The total simulated time for which the KFWH solution was computed ranged from 28 ms to 45 ms, depending on the particular case. The simulation employed a sampling time step of 11 μs, resulting in a Nyquist frequency of approximately 45 kHz. Due to the relatively short sampling lengths window averaging was not feasible, and third-octave-band-averaged sound levels therefore are presented herein.

Third-octave band levels from the experimental measurements are compared with values obtained from the URANS solutions by direct pressure sampling and by employing the KFWH analogy for the ten experimental conditions in Figure 2.10. The good agreement between the numerical solutions that can be seen in the plots of Figure 2.10 is expected, because the hydrophone probe was located within the acoustic nearfield of the KFWH model. Compared to the experimental measurements, the overall sound pressure levels were over-predicted by simulations by 20 dB to 60 dB. However, the shapes of spectra were generally captured well, and the difference between the experimental and the numerical values was generally consistent across all frequency bands. Significant mismatch in the spectral features occurred when the numerical simulations failed to capture the dominant cavitation structures. For example, the numerically predicted spectra corresponding to the conditions C2 and C5 lack the characteristic hump associated with tip vortex cavitation that can be clearly seen in the experimental data. Also, the shape of the spectrum corresponding to the condition C3, which exhibited bubble cavitation, did not match the experimental data trend at high frequencies, which was expected given that bubble cavitation was not captured numerically. Good reproduction of the spectral shapes was seen for the conditions under which pressure side sheet cavitation, was accurately reproduced by the hydrodynamic model, was produced (C4 and C1b through C4b). The good match of the spectral patterns when cavitation patterns matched, as well as the lack of spectral features associated with phenomenon that were not captured by the simulations, indicate good performance of the acoustic models in terms of spectral acoustic power distribution.

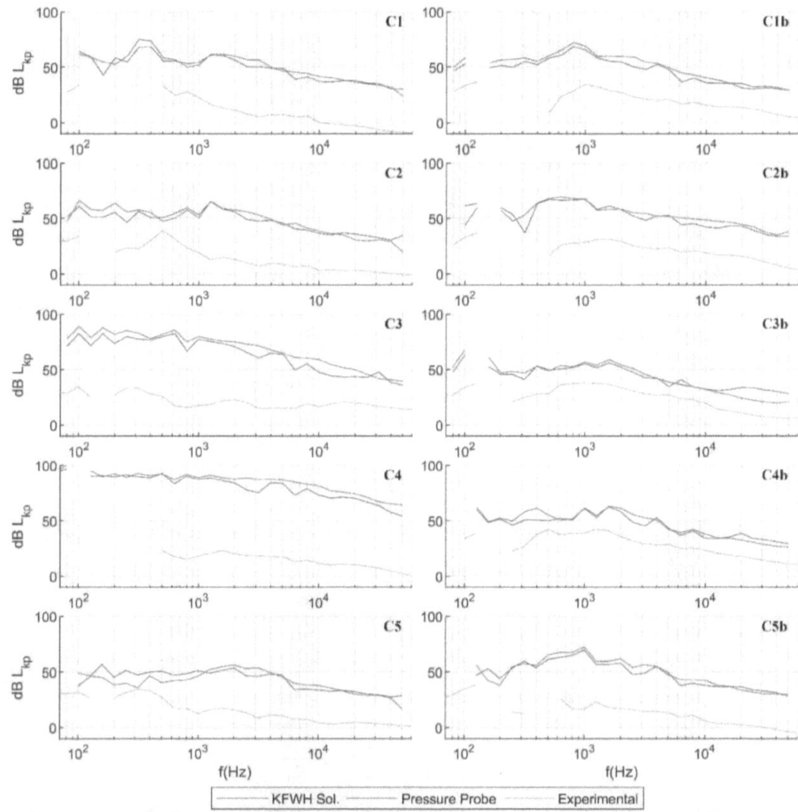

Figure 2.10: Third-octave band level spectra obtained by experimental measurements, direct pressure calculations and KFWH hydroacoustic model solutions. Left column: design pitch operation. Right column: reduced pitch operation.

The over-prediction of the sound pressure levels indicates that significant damping effects were neglected in the present modelling procedure. Neglected flow-sound integrations and wall vibrations are unlikely to account for an over-prediction of pressure fluctuations by multiple orders of magnitude that was observed in the simulations. Therefore, the most likely source of lost damping effects in the present modelling procedure are the shortcomings of the Schnerr and Sauer cavitation model that governs interphasic mass transfer. The model neglects several important sources of damping in the oscillation of the bubbles, including thermal, viscous, and acoustic radiation damping. These limitations are discussed in the following section.

2.5.6 Damping model

The relative influence of different types of damping on the oscillation of cavitation bubbles, and in turn on the radiated sound, is dependent on the size of the bubbles. In order to make order of magnitude estimates of the importance of different damping types on the growth and collapse of a bubble, it is convenient to assume that cavitation bubbles oscillate linearly and compute their sizes accordingly. The oscillation frequency is then a function of the equilibrium bubble radius R_0, surface tension γ_S, and equilibrium partial pressure of gas p_{g0} in the bubble:

$$f_0 = \frac{1}{2\pi R_0} \sqrt{3\gamma p_{g0} - \frac{2\gamma_S}{R_0}}. \tag{2.16}$$

The latter can be found from the ambient liquid and the saturated vapour pressures and the equilibrium bubble radius [31]:

$$p_{g0} = p_\infty + \frac{2\gamma_S}{R_0} - p_v. \tag{2.17}$$

Given the range of ambient pressures and acoustic frequencies of interest, the sizes of bubbles can be estimated according to the following expression:

$$f_0 = \frac{1}{2\pi R_0} \sqrt{3\gamma \left(p_\infty + \frac{2\gamma_S}{R_0} - p_v \right) - \frac{2\gamma_S}{R_0}}. \tag{2.18}$$

In the present work, the expected range of ambient equilibrium pressures can be assumed to be bounded by the vapour pressure of water and the maximum pressure in the cavitation tunnel. Limiting the range of frequencies to those between 100 Hz and 100 kHz, where substantial cavitation noise was observed in the experiments, and assuming diatomic gas ($\gamma = 1.4$) gives a range of bubble sizes from centimetres down to tens of micrometres. At these scales, the viscous damping term in the Rayleigh-Plesset equation is small. Introducing these values into a linearized form of the equation results in viscous damping ratios of the order of 10^{-6} to 10^{-7}, suggesting that viscous damping is not significant to the bubble motion.

In fact, oscillation damping in primarily gas-filled spherical bubbles that range between hundreds of millimetres and tens of micrometres in radius is dominated by thermal damping resulting from polytropic expansion and compression. Acoustic radiation-related damping, which dominates at larger scales, plays a significant secondary role [32], [62]. While these two sources of damping are not easily quantified in the context of the present results, the findings nonetheless point to them as important factors for numerical prediction of cavitation noise.

The lack of adequate damping in the present numerical model represents a significant deficiency that must be addressed before the methodology is widely applicable. While the Schnerr and Sauer cavitation model has been successfully used in the past studies to predict cavitation patterns, the present results suggest that it is inadequate for prediction of the radiated sound. An investigation of the influence of the cavitation models in numerical prediction of the radiated noise from cavitating propellers, with a focus on damping, should be carried out as part of the future work.

2.6 Conclusions

We performed model-scale experiments involving measurements of the radiated acoustic noise emitted by a cavitating propeller. The results were replicated numerically using a CFD method based on solution of the URANS equations in conjunction with an acoustic analogy. The experimental results provided insight into the relationships between the predominant cavitation regimes and their acoustic signatures. The capabilities of common numerical modelling strategies with regards to specific features of propeller cavitation were assessed in the course of the numerical study.

The experimental results highlighted the key acoustic features related to the four observed cavitation regimes, namely the narrow peak associated with TVC, broader peak of TVC with sheet cavitation, the irregular high-frequency behaviour of bubble cavitation, and the broad decaying spectrum of pressure side cavitation. Results indicated that a small number of high-level parameters may be sufficient for developing a semi-empirical modelling strategy. The relatively high sound levels associated with pressure-side sheet cavitation are noteworthy in that their association with reduced-pitch and low-load operation indicates that reduction of the vessel's speed is not a sufficient condition for quiet operation. The present results highlight a need for intelligent strategies for mitigation of the cavitation noise, particularly in the context of speed reduction in vessels with controllable pitch propellers.

The present numerical approach accurately predicted the cavitation patterns, as well as the distribution of the acoustic power through the frequency spectrum, in cases where sheet cavitation was observed in isolation. A URANS formulation is not sufficient for prediction of cavities within the tip vortices due to numerical damping of the vortices themselves. The Schnerr and Sauer cavitation model over-predicted the levels of the radiated sound, suggesting the need for an improved cavitation model that would include sound damping due to thermal effects and acoustic radiation.

Chapter 3: Parametric Mapping Approach for Cavitation-Induced Acoustic Emissions from Marine Propellers

Duncan McIntyre [a], Zuomin Dong [a], Giorgio Tani [b], Fabiana Miglianti [b],
Michele Viviani [b], Pengfei Liu [c], Peter Oshkai [a]

[a] Department of Mechanical Engineering, University of Victoria, Canada
[b] Department of Naval, Electrical, Electronic and Telecommunications Engineering,
University of Genoa, Italy
[c] School of Engineering, Newcastle University, UK

3.2 Introduction

Climate change has brought widespread attention to the need for mitigation of chemical emissions across industries, including the marine transportation industry. Chemical emissions mitigation has been widely accepted as an objective in the optimization of ship design and operation. However, ships also represent the world's largest source of underwater noise pollution [17]. Marine fauna rely heavily on sound for communication and sensing, and underwater radiated noise (URN) level is a critical factor affecting the health of marine ecosystems [6], [8], [10]. Chemical emissions are directly related to engine operation, and prediction techniques for a variety of chemical emission types are well-established in both academic literature and industry. In contrast, prediction of acoustic emissions based on vessel operating conditions remains challenging. URN from ships is dominated by propeller-induced cavitation [18]. The cavitation phenomenon is the process of vaporization that occurs when a liquid is brought below its local critical pressure, which is commonly associated with flow structures generated by marine propellers. Depending on the propeller and vessel operating conditions, one or more different cavitation regimes may be induced by the propeller, resulting in variable radiated acoustic energy and distinct spectral characteristics [50], [63]. The task of predicting radiated noise from ships is therefore reliant on modelling procedures for connecting vessel operating conditions to the emission of sound from propeller-induced cavitation. The present work aims to establish a framework for prediction of acoustic emissions in a way that allows simultaneous consideration of chemical and acoustic emissions for optimization of marine vessel operation. An example map based on model-scale experimental measurements is generated as a proof-of-concept.

Chemical emissions from combustion engines vary according to the operating conditions such as fuel consumption and combustion temperature. Comprehensive tools for predicting these emissions have been developed to pair with regulations and are now mature and widely implemented, in various forms, globally. Brake-specific emissions maps are a common and convenient way to present engine emissions characteristics, as they allow immediate estimation of a specific emission type from operating conditions that are usually known, e.g., engine speed and torque. An example of a set of maps of this type for the DAF FT CF75 tractor, a commercial road transport vehicle powered by a PACCAR PR model 9.2 litre diesel cycle internal combustion engine, is shown in Figure 3.1. The present work aims to construct similar maps for the URN emissions, allowing convenient evaluation of chemical and acoustic emissions from marine vessels in the context of optimization.

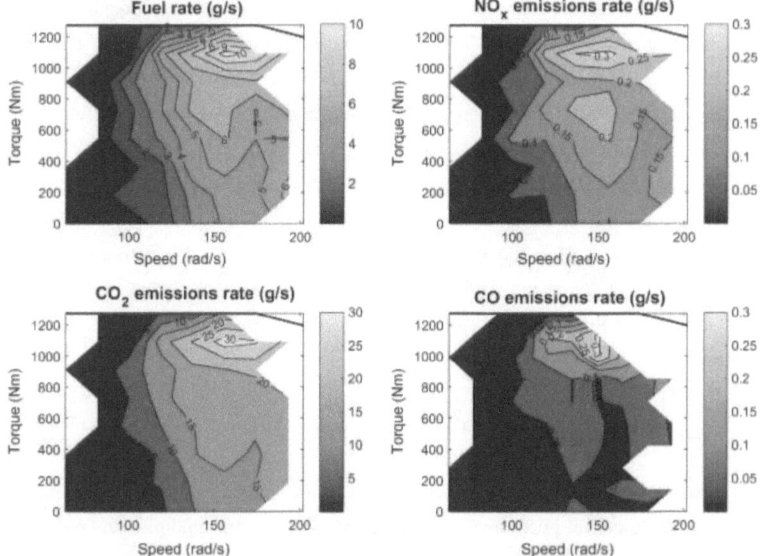

Figure 3.1: Contour plots of transient brake-specific levels of fuel consumption and engine-out emissions of NOx, CO_2 and CO as functions created using the optimum bins for the DAF CF75 tractor [64].

Models for the prediction of propeller noise have existed since the Second World War [18], but the physical mechanisms governing the production of sound by propeller-induced cavitation are not well understood. Early URN models relied primarily on ship speed as a predictive tool, ignoring the significant influence of variation in the types of propeller cavitation and the dynamic effects of propeller shape on cavitation production. Later, semi-empirical models, such as Brown's formulation [65], were designed to relate radiated noise to the amount of cavitation induced by the ship's propeller. Attention to the issue of URN from shipping has grown in recent years, resulting in increased research interest in physics-based models of cavitation noise. Reliable boundary element method (BEM) techniques have been established for the prediction of vapour cavities and the resulting loading on marine propellers since at least the mid-2000s (e.g., [35]). Boundary element methods do not resolve the flow field, and do not provide information about the three-dimensional evolution of vapour cavities that is required to compute an acoustic solution. Therefore, emitted noise levels must be computed by combining the BEM with a semi-empirical model. Significant advances in comprehensive finite-volume techniques for cavitating propeller flows, including Reynolds-Averaged Navier-Stokes (RANS) (e.g., [23], [25], [36], [66], [67]) and Large Eddy Simulation (LES) (e.g., [38]), have been made over the past decade. The addition of acoustic considerations to computational fluid models has also been attempted by a several research groups [40], [41], [63], [68]. While finite-volume methods have the potential to provide greater physical insight and generality compared to combined BEM/semi-empirical models, the large increase in computational expense (for example, computing the RANS-based solutions presented in Chapter 2 required tens of thousands of compute core-hours per simulation, while the BEM results presented in this chapter required only a few core-minutes) makes lower-order models attractive for optimization applications.

3.3 Mapping Procedure

Mapping cavitation-induced on an operating parameter space requires prediction of multiple physical phenomena, including propeller hydrodynamics and cavitation acoustics. The proposed methodology uses a series of numerical and empirical models to relate the rotation speed and torque output of a marine propeller to the level of cavitation-induced noise generated at the propeller. Figure 3.2 shows a flowchart that maps the operating parameters and flow of information needed to connect engine operation with cavitation-endued noise. Three interconnected models are needed, representing the flow around the propeller, the dynamics of the vapor cavities, and the acoustics. The inflow conditions to the propeller are taken as an input in the present modelling process. In practice, a hydrodynamic model of the hull would be required. The propeller inflow can be determined by model-scale testing in a towing tank [52] or by using numerical techniques such as panel methods or RANS solutions [69]. Model-scale measurements used as a proof of concept in the present work are based on uniform inflow, and wake structure of the hull was not considered.

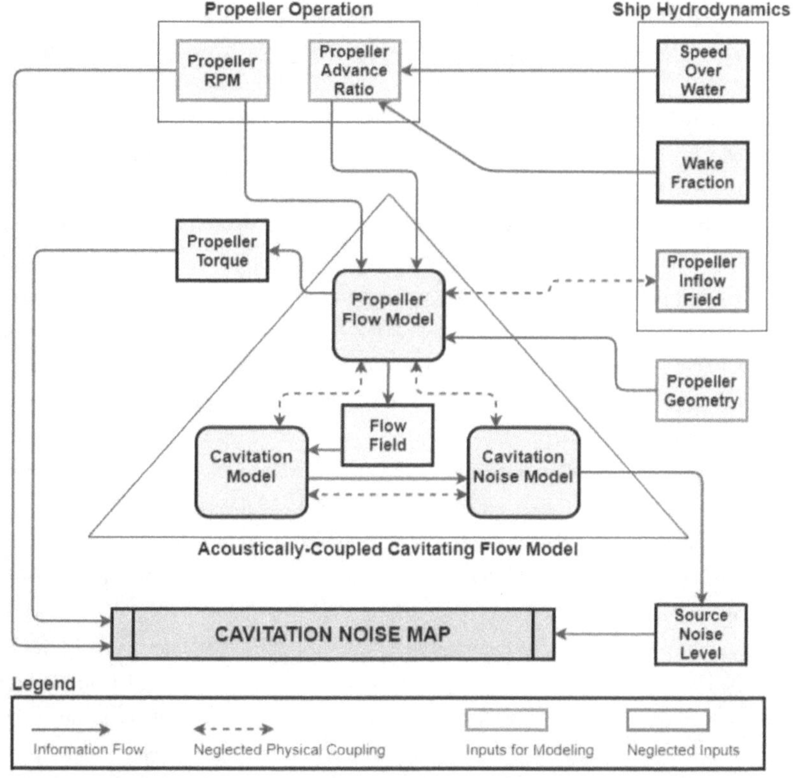

Figure 3.2: Information flow in the proposed mapping methodology

Noise generated by ship propellers is greatly increased by the presence of cavitation [18], [50], [70]. As the flow around a marine propeller is accelerated, localized drops in pressure can result in formation of vapour cavities within seawater. Growth, oscillation, and collapse of these cavities cause high-amplitude fluctuations in the local pressure, which are radiated into the acoustic far-field. Modelling of the cavitating flow around the propeller and the coupled noise model are the aspects of the mapping procedure that represent the main focus of the present work. The flow field, the phase change, and the acoustic propagation are mutually coupled, and each coupling is bidirectional. Cavitating flow over a propeller is both multiphase and multiscale, making it difficult to accurately reproduce using CFD techniques; however, some success has been achieved with both

Unsteady Reynolds Averaged Navier-Stokes (URANS) (e.g., [23]–[25], [27], [36]) and LES (e,g., [38]) solutions. While sound propagation through fluids can be represented using vortex sound theory [71], both URANS and LES do not account for the turbulence scales required to model noise propagation. Therefore, an additional, coupled noise model must be used in conjunction with the hydrodynamic equations in order to model the far-field acoustic effects ([39]–[42].)

In the application of acoustic emission mapping, computational expense is of primary importance when selecting modelling tools. While high-fidelity CFD techniques can be used for a small number of simulations, creating a single emissions map requires tens or hundreds of simulations of the cavitating flow field in order to represent the necessary number of operating points. The present mapping methodology therefore relies on a low order source-doublet panel method solver for flow prediction. Specifically, the proof of concept provided herein employs the panel solver ROTORYSICS [72], one example of such a code. Potential flow solvers have been shown to provide accurate predictions of thrust and torque for propellers experiencing cavitation [73]–[75]. These solvers can be combined with cavitation models that use the critical pressure threshold to predict sheet cavitation on the surface of propeller blades. However, these methods are incapable of capturing tip vortex cavitation.

Since direct noise modelling based on acoustic pressure fluctuation in the flow is not possible using a low-order panel method, semi-empirical noise modelling has been used in the present methodology. The most popular example of such is a model is Brown's formulation, which uses the area of sheet cavitation on a propeller blade to predict the source noise levels as a logarithmically decaying spectrum [76]. Empirical models for tip vortex cavitation-induced noise have also been developed based on the use of circulation to estimate the strength of the tip vortices [77].

Combining a panel method flow solver, a critical pressure cavitation model, and a semi-empirical noise model satisfies the mapping requirements shown in Figure 3.2. This approach is computationally efficient, allowing both propeller toque and noise to be predicted for a given set of operating conditions. The present methodology is also customizable. For example, URANS solution may be used in place of the panel method for more accurate modelling of critical operating conditions without changing the overall methodology of emissions mapping.

3.4   Methodology

3.4.1   Experimental system and techniques

The experiments presented in Chapter 3 were used for numerical benchmarking in the present work. A brief description of the experimental system and results are provided herein for clarity.

3.4.1.1 Experimental setup

Model-scale propeller tests were conducted in a closed-circuit cavitation tunnel at the University of Genoa. The tunnel configuration, illustrated schematically in Figure 2.2, generated uniform inflow with a maximum velocity of 8.5 m/s at the entrance of a 2.2-m-long test section that had a cross-section of 0.57 m x 0.57 m. We maintained the concentration of dissolved oxygen at 4.5 ppm, which was found to provide sufficient seeding for cavitation while simultaneously minimizing noise absorption by free bubbles. The model-scale propeller was positioned near the inlet of the test section with its supporting pod located downstream, as shown in Figure 2.3, to ensure uniform inflow conditions.

The propeller used in the experiments was a scale model of a four-bladed controllable-pitch propeller (CPP) representative of a mid-size tanker vessel. The parameters of the propeller are presented in Table 2.1. We studied two pitch configurations: the designed pitch ($[P/D]_{0.7R} = 0.87$) and a reduced pitch ($[P/D]_{0.7R} = 0.521$). A Kempf & Remmers H39 dynamometer measured the rotation speed, thrust, and torque of the propeller. The corrections of Wood and Harris [51] were used to correct the dynamometer measurements for tunnel effects. Three Vision Tech Marlin F145B2 Firewire cameras captured images of the cavitation with illumination provided by two stroboscopic lights (900 Movistrob). We obtained three concurrent views of the propeller: a view of the pressure side, of the suction side of a single blade, as well as a view of the entire propeller.

Two active hydrophones, a Bruel & Kjaer type 8103 (H_1) and a Reson TC4013 (H_2), captured acoustic information. The hydrophone H_1 was submerged in water and separated from the test section by a plexiglass window, while hydrophone H_2 was located inside the test section as shown in Figure 2.3. Post-processing of acoustic measurements was conducted according to the ITTC guidelines for model-scale noise measurements [52], including background noise correction. In this paper, we present the power spectra in terms of the non-dimensional pressure coefficient K_P, described by:

$$L_{KP}(f) = 20 \log_{10} \left(\frac{K_P(f)}{10^{-6}} \right), \quad K_P = \frac{p}{\rho n^2 D^2}, \tag{3.1}$$

where p is the local pressure, ρ is the liquid density, n is the revolution rate of the propeller (in revolutions per second), and D is the propeller diameter.

3.4.1.2 Operating conditions and their corresponding cavitation regimes

Ten propeller loading conditions, five each for the design- and reduced-pitch settings, were tested. We varied the advance ratio and the cavitation number to induce each of the cavitation regimes shown in Figure 2.1. Tip vortex cavitation, bubble cavitation, and sheet cavitation on the suction and pressure sides of the propeller blade were all observed in the experiments. Among these cavitation regimes, tip vortex cavitation occurs at the highest nominal cavitation number and produces the lowest overall level of noise. The vortex cavity itself exists within the core of the tip vortex generated by the blade and can therefore persist for many diameters downstream of the propeller. Sheet cavitation most commonly appears on the suction side of the propeller blade near the tip, but may also occur on the pressure side when the propeller is under-loaded. Pressure side cavitation may also include vortex-from-sheet or leading edge vortex cavities, but these regimes have been considered as a unified group in the present work. Bubble cavitation occurs at the lowest cavitation numbers and appears most commonly of the suction side around mid-span of the blade [50].

The experimental conditions and the corresponding cavitation regimes are summarized in Table 2.2.

3.4.2 Computational model

3.4.2.1 Panel method and influence coefficients

In the present investigation we employed a BEM to solve the potential flow equation in order to acquire the hydrodynamic inputs for necessary for modelling cavitation-induced noise. The software package ROTORYSICS was employed to run the panel method simulations. In the present method, a surface mesh of polygonal panels formed the boundary where the potential flow equations were discretized. Each panel was assumed to have constant doublet and source strength distributions at a given time step. The mesh included the surfaces of the propeller hub, the blades, wake bodies used to simulate the propeller wake. Source and doublet strengths were solved from the potential flow equations iterative for the centroids of each panel. Differentiation of the panel potentials according to the unsteady Bernoulli equation yielded the velocities and the pressure distribution over the surfaces [73]. A mathematic description of the panel method is presented in Appendix B.

3.4.2.2 Cavitation modelling with a critical pressure scheme

Analytical descriptions of volume dynamics and acoustics of oscillating cavitation bubbles exist for spherical bubbles in infinite media [19]. Moreover, volume-of-fluid methods based on spherical bubble models, have been successfully used in conjunction with in RANS and LES CFD simulations [21], [28]. In potential flow solvers, however, a critical pressure scheme is more commonly implemented. The scheme used in the present work is given in [73]. The model is based on the local cavitation number and the local pressure coefficient, where the local pressure coefficient was defined as:

$$C_P = \frac{p - p_{\text{ref}}}{\frac{1}{2}\rho U_\infty^2},$$

(3.2)

and the local cavitation number was defined as:

$$\sigma = \frac{p_{\text{ref}} - p_V}{\frac{1}{2}\rho U_\infty^2},$$

(3.3)

where U_∞ is the free-stream velocity, p is the local pressure, p_{ref} is the local hydrostatic reference pressure, and p_V is the vapour pressure of the fluid. When the value of C_P was below $-\sigma$, which corresponded to the local pressure being below the vapour pressure, the cavitation was assumed to occur at the given location in the flow field. In the context of propellers, a nominal cavitation index is conventionally used:

$$\sigma_N = \frac{p_{\text{ref}} - p_V}{\frac{1}{2}\rho n^2 D^2},$$

(3.4)

where p_{ref} is taken to be the hydrostatic pressure at the propeller shaft depth. Thus, conversion from the nominal to the local cavitation number was necessary for predicting inception of cavitation:

$$\sigma = \frac{\frac{1}{2}n^2 D^2 \sigma_N - gz'}{\frac{1}{2}U_\infty^2},$$

(3.5)

where z' is vertical coordinate relative to the centre of the propeller shaft. Once cavitation inception was determined, the extent of the cavity was predicted using a method illustrated in Figure 3.3.

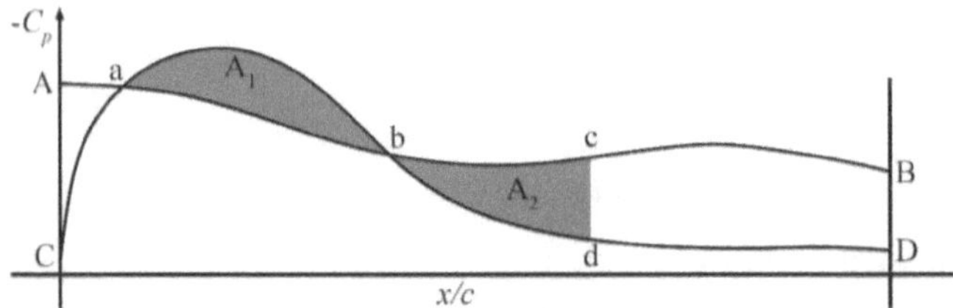

Figure 3.3: Illustration of the critical pressure scheme for prediction of the extent of cavitation along the chord length (c) of the propeller blade. x represents the chordwise coordinate.

In Figure 3.3, the curve C-D represents the distribution of the pressure coefficient along the chord of the propeller blade, as predicted by potential flow theory. The curve A-B represents the distribution of the local cavitation number. Area A_1 is the "chop-off area," where the pressure is below the vapour pressure; area A_2 is the "fill-in area," and is equal to A_1. At all chordwise positions between points a and c, the pressure is assumed to the equal to the critical pressure:

$$C_p = -\sigma, \tag{3.6}$$

i.e. the pressure distribution along the chord is given by curve A-B between points a and c and by curve C-D elsewhere. This procedure corresponds to the assumption that the cavity remained attached to the blade, that its growth occurs in the range of chordwise positions between points a and b, and that is collapses between points b and c.

In the present work, the model described above was modified to take into account the results of the model-scale experiments from Chapter 2. The critical pressure scheme in conjunction with the source-doublet potential flow model over-predicted the pressure drop at the mid-chord of the propeller blade, resulting in over-prediction of the extent of cavitation at that location. During the experiments, cavity formation for all regimes, except bubble cavitation, occurred at or near the leading edge of the propeller blade. Therefore, in the numerical simulations, cavities that did not form within 10% of the total chord from the leading edge of the propeller blades were neglected.

3.4.3 Noise Modelling

3.4.3.1 Semi-empirical noise model

Noise modelling of cavitation patterns predicted from numerical results have been predicted successfully in literature using semi-empirical noise models [68]. One of the most widely used of these is Brown's formulation, intended for sheet cavitation [76]:

$$L_P = 163 + 10\log_{10}\left(\frac{ZD^4n^3}{f^2}\right) + 10\log_{10}\left(\frac{A_C}{A_D}\right).$$ (3.7)

In Brown's formula, sound pressure level (L_P, in dB with a 1 µPa reference), is estimated based on the number of propeller blades Z, the propeller diameter, the rotation speed, the frequency of interest f, and the proportion of the two-dimensional disc area swept by propeller over which cavitation occurs A_c/A_D. A_c is a two-dimensional projection of the cavitation pattern on the surface of the propeller onto a plane orthogonal to the propeller's axis of rotation, and A_D is the swept disc area of the propeller blades. The quantity A_c/A_D can be obtained easily from a panel method simulation that includes cavitation.

More recently, Bosschers presented the second version of the Empirical Tip Votex method (ETV-2) [77]. This model uses circulation close to the propeller blade's tip to the estimate the radiated noise spectrum from the cavitation in the tip vortex. The maximum sound pressure level $L_{KP,\mathrm{max}}$ and centre frequency f_c of radiated noise are estimated by two semi-empirical formulae:

$$L_{KP,\mathrm{max}} = a_p + 20\log_{10}\left\{\left(\frac{r_c}{D}\right)^{k_r}\sqrt{Z}\right\},$$ (3.8)

$$\frac{f_c}{f_{bpf}} = b_f\,\frac{1}{r_c/D}\,\frac{\sqrt{\sigma_n}}{Z}.$$ (3.9)

where r_c is the radius the tip vortex cavity and $f_{bpf} = Zn$ is the blade passing frequency. The constants a_p, b_f, and k_r are determined empirically, while the radius of cavitation within the tip vortex is taken from a combination of a critical pressure assumption and a vortex model. In the present work, an inviscid vortex model is used for the sake of simplicity. For an inviscid cavitating vortex, the cavitation radius takes a closed form in terms of circulation Γ_∞, propeller rotation speed, and nominal cavitation index.

$$r_c = \frac{1}{2\pi}\frac{\Gamma_\infty}{n}\frac{1}{\sqrt{\sigma_n}}.$$ (3.10)

Bosschers recommends the circulation about the blade at 95% of the propeller radius be used for modelling [77]. Circulation can be obtained readily from panel method simulation results.

The shape of the power spectrum is estimated in the ETV-2 model by two empirical equations, the first representing the characteristic hump of tip vortex cavitation and the second representing the decay spectrum common to most cavitation noise:

$$H_h(f) = 20 \log_{10} \left\{ \operatorname{sinc} \left(\frac{f - f_c}{0.830 \Delta f_{-6\text{dB}}} \right) \right\}, \tag{3.11}$$

$$H_s(f) = 10 \log_{10} \left\{ \frac{2 \left(\frac{f}{f_c} \right)^{\alpha_c}}{1 + \left(\frac{f}{f_c} \right)^{\alpha_l - \alpha_h}} \right\}. \tag{3.12}$$

where $\Delta f_{-6\text{dB}}$ is the width of the hump at half of its peak magnitude. The shape parameters in these functions, α_l, α_c, and α_f, are all determined empirically. The final shape of the spectrum is then computed by combining the two spectral shapes with a weighting factor α_0.

$$H(f) = 10 \log_{10} \left\{ \alpha_0 10^{\frac{H_h(f)}{10}} + (1 - \alpha_0) 10^{\frac{H_s(f)}{10}} \right\}, \tag{3.13}$$

$$L_{KP}(f) = L_{KP,\max} + H(f). \tag{3.14}$$

The present work combined a modification of Brown's formulation with the ETV-2 model in order to produce a formula applicable to a wider range of cavitation regimes. The empirical constants were chosen based on the model-scale experiments discussed in Chapter 2, and are only valid for that specific propeller; further work is required to implement and validate a similar model at full-scale. Thus, the present empirical noise model is only a proof-of-concept and not a validated model.

Three sound pressure levels were added in the present model: two spectra calculated by modifications of Brown's formulation to represent sheet cavitation on the suction sides of the propeller blades (denoted by the P and S subscripts), and one calculated using the ETV-2 model (denoted by the ETV subscript):

$$L_{KP}(f) = 10 \log \left\{ \beta_S 10^{\frac{[L_{KP}(f)]_S}{10}} + \beta_P 10^{\frac{[L_{KP}(f)]_P}{10}} + \beta_{ETV} 10^{\frac{[L_{KP}(f)]_{ETV}}{10}} \right\}. \tag{3.15}$$

where the various subscripts of β represent the weighting factors for the spectra corresponding to the three regimes of cavitation. The approximation of the three component spectra relied on a maximum value alongside a shape function. The formulae for estimating the maximum level and shape function for the ETV spectrum were unchanged from the published version of the ETV-2 model. Noise induced by sheet cavitation lacks the tonal component represented by the shape function H_h in the ETV-2, and the form of shape function H for the spectra representing these noise sources therefore took the same form as the function H_s in the ETV-2 model. For the suction side spectrum:

$$[H(f)]_S = \left[10 \log_{10}\left\{\frac{2\left(\frac{f}{f_c}\right)^{\alpha_c}}{1 + \left(\frac{f}{f_c}\right)^{\alpha_l - \alpha_h}}\right\}\right]_S . \tag{3.16}$$

The form of H was identical for the pressure side spectrum, although the value of the shape constants and centre frequency were different. A modification of Brown's formula gave the maximum level. For the suction side:

$$\left[L_{KP,\max}\right]_S = \left[a_p\right]_S + 10 \log_{10}\left(\frac{ZD^4 n^3}{[f_c^2]_S}\right) + 10 \log_{10}\left(\frac{[A_c]_S}{A_D}\right), \tag{3.17}$$

and likewise for the pressure side. Each of the constants a_p had to be determined empirically.

Cavitation patterns computed with the panel method determined the values of the influence coefficients β_S, β_P, and β_{ETV} according to criteria determined *a priori*. Suction side cavitation was exclusively observed in isolation in the experiments; therefore, the influence coefficient vector $[\beta_S\ \beta_P\ \beta_{ETV}]$ was defined to be $[0\ 1\ 0]$ in the case that cavitation formed on the pressure side of the propeller blades. In the case that pressure side cavitation covered an area of less than 0.1% of the propeller's swept area, a combination of suction side sheet and tip vortex cavitation was assumed. An empirical equation was used to determine the values of β_S based on the area of cavitation on the suction side of the blade:

$$\beta_S = C_S \left\{\frac{[A_c]_S}{A_D}\right\}^{k_S} . \tag{3.18}$$

where C_S and k_S were empirical constants. The influence coefficient vector takes took form $[\beta_S\ 0\ (1 - \beta_S)]$ when a combination of tip vortex and suction side sheet cavitation appears in the simulation, with the limiting case of $[0\ 0\ 1]$ arising in the absence of cavitation on the surface of the blades. Other forms of cavitation, including bubble and hub cavitation, could not be accounted for in the present hybrid model.

3.4.3.2 Application of noise models to simulation results

Each panel method simulation ran for three propeller revolutions with a time step equal to $1/36^{\text{th}}$ of a revolution period. Final results are presented as averaged values from the 36 time steps of the final revolution. When interpreting cavitation patterns as areas for noise modelling, the area of each panel was scaled by the proportion of the time steps in one revolution for which cavitation was present on that panel. Cycle averaged values were used to calculate cavitation area and circulation, the inputs for semi-empirical noise modelling. Radiated noise level spectra were then obtained according to Eqn. (3.15)

In order interpret the power spectra output from semi-empirical noise models, we applied a weighting function and performed a trapezoidal integration of the weighted spectra. We selected the M-weighting function for high-frequency cetaceans proposed in [78] for its applicability to local southern resident killer whale populations in the vicinity of Victoria, Canada for the proof-of-concept mapping procedure. In principal, other weighting functions could be applied in the same way; however, it is worth noting that the applicability of semi-empirical cavitation noise models to lower-frequency noise from ships might be limited both by an increased complexity of the acoustic signature and the increased importance of other noise sources in that frequency range.

3.5 Results and Discussion

3.5.1 Overview of cavitation regimes and corresponding noise characteristics

3.5.1.1 Cavitation patterns from numerical simulation

Ten numerical simulations reproduced the ten conditions from the experimental campaign discussed in Chapter 2 in terms of pitch, cavitation index, and propeller advance ratio. Figure 3.4 presents a comparison of the suction side cavitation patterns collected in the experimental campaign to numerical results from the panel method simulations with and without applying the empirical correction designed to correct for the over-prediction of cavitation at mid-blade. The same comparison for pressure side cavitation patterns is given for designed pitch conditions and reduced pitch conditions in Figure 3.5 and Figure 3.6 respectively. Without the correction, a cavity appears in the majority of the simulations at mid-blade that was not observed in the experiments. Based on this comparison, we correct the cavitation patterns by neglecting any cavity that does not arise within 10% of the chord from the leading edge of the blade. In experiments, only bubble cavitation was found to appear mid-blade.

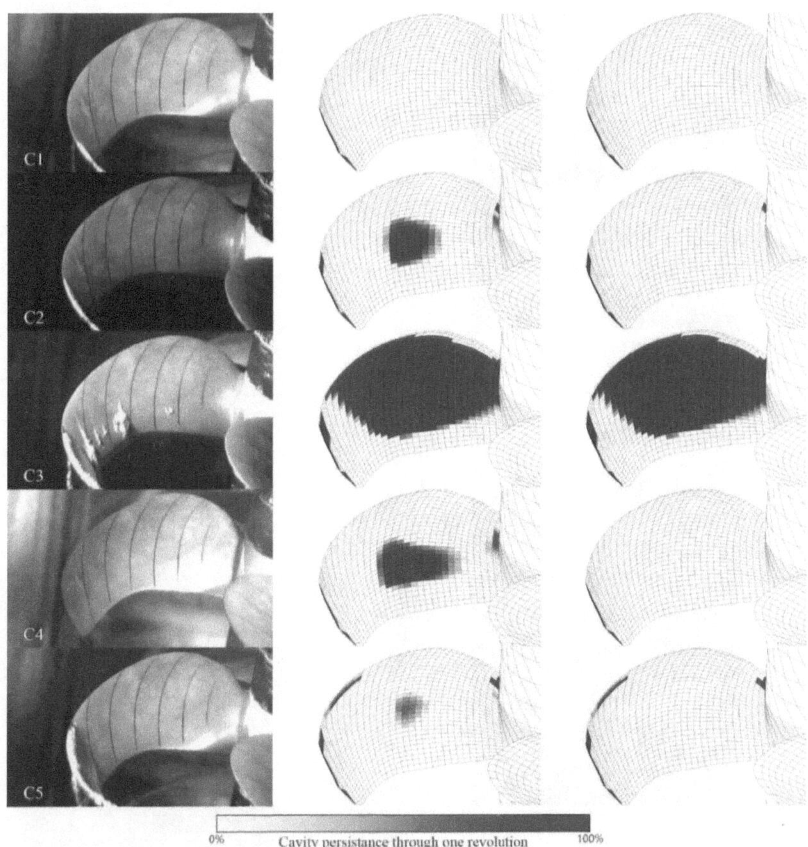

Figure 3.4: Suction side cavitation patterns induced by design-pitch operation of the model propeller obtained from experiments and simulations. Left column: stroboscopic photographs. Centre column: panel method results. Right column: empirically corrected panel method results.

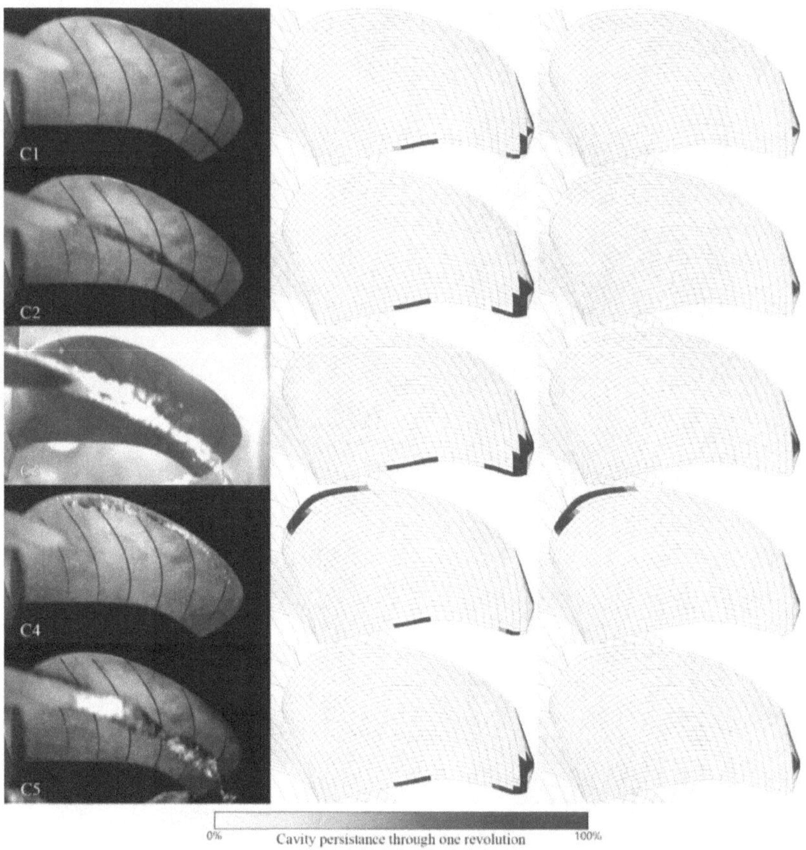

Figure 3.5: Pressure side cavitation patterns induced by design-pitch operation of the model propeller obtained from experiments and simulations. Left column: stroboscopic photographs. Centre column: panel method results. Right column: empirically corrected panel method results.

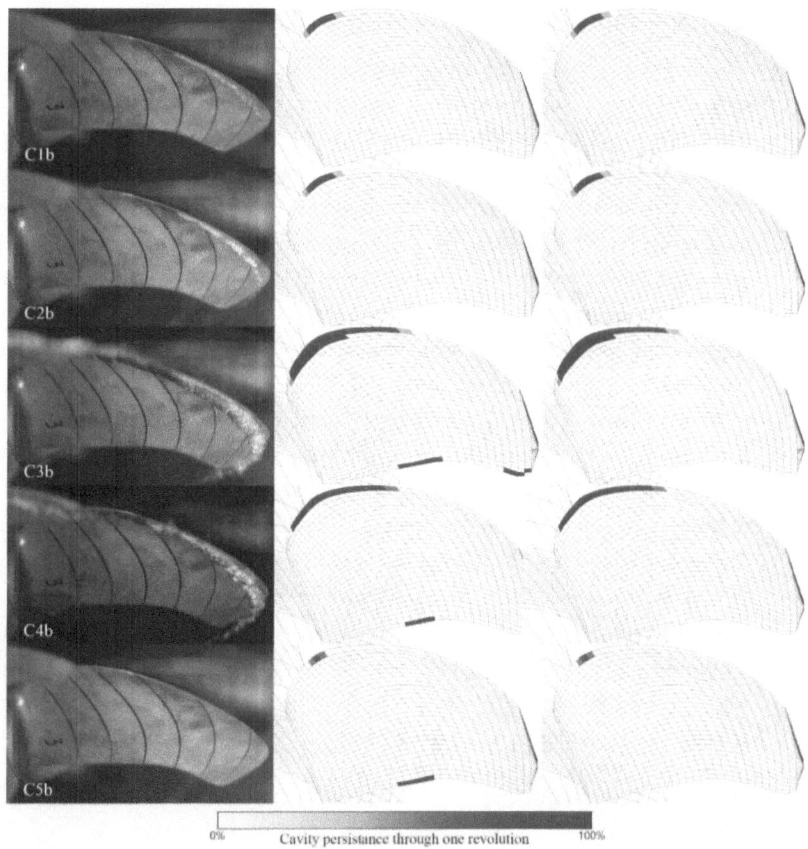

Figure 3.6: Pressure side cavitation patterns induced by reduced-pitch operation of the model propeller obtained from experiments and simulations. Left column: stroboscopic photographs. Centre column: panel method results. Right column: empirically corrected panel method results.

Panel method simulations accurately predicted the type of cavitation, showing small cavities at the blade tip in conditions that produced tip vortex cavities and cavities at the leading edge of the pressure side in conditions that produced pressure side cavitation. The extent and exact locations of individual sheet cavities were not accurately predicted, although the relative sizes of cavities appeared qualitatively correct when comparing different conditions.

3.5.1.2 Semi-empirical noise models

We computed acoustic signatures from the panel method simulation results replicating each of the experimental conditions using three semi-empirical noise models: Brown's formula [76], the ETV-2 model [77], and the new hybrid model. The coefficients for Brown's formula are taken directly from [76], while the empirical coefficients for the other two models were chosen in order to fit the experimental data. As the propeller used in the present work was a scale model, the empirical constants are not valid for full-scale propellers; however, the results reveal the types of characteristics than can be captured by each modelling methodology. Figure 3.7 presents the results of each of the three modelling strategies in comparison with experimental measurements from the ten experimental conditions.

Brown's formulation gives uniformly decaying spectra, which limits its range of potential applicability to the high-frequency range. The relative magnitude of noise levels predicted by Brown's formulation do not match experimental results, especially when used to predict the noise from pressure side cavitation compared to other cavitation regimes. The difference between the performance of the ETV-2 and hybrid models was not substantial for conditions that were dominated by tip vortex cavitation. The largest difference appeared at the high end of the measured frequency rage, where the change in decay rate was captured marginally better by the hybrid model compared to the ETV-2 model. The ETV-2 did not predict the levels or spectral shapes of the noise generated by pressure side cavitation, which was expected.

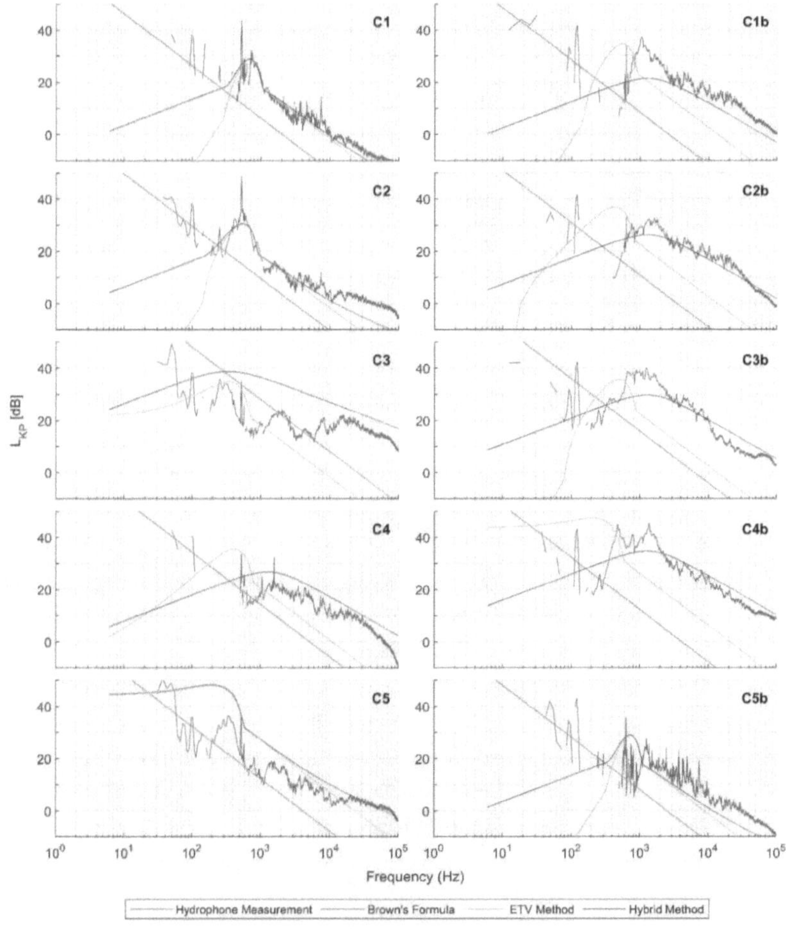

Figure 3.7: Comparison of cavitation noise power spectra measured in experiments via hydrophone against predictions using the panel method alongside Brown's formula, the ETV-2 model, and a hybrid acoustic model.

When considering pressure side cavitation, the hybrid model was solely capable of providing an acceptable approximation; this result was expected, as it the hybrid model was the only formulation designed to be used for the pressure side regime. The present result highlights the primary advantage of a hybrid modelling strategy, in that it has the potential to account for cavitation regime transitions when considering a wide operating regime space in the context of optimization.

3.6   Noise emission maps

We applied the mapping procedure using panel method simulations and the hybrid empirical cavitation noise model to generate a map of cavitation-induced noise, on an RPM-torque parameter space, for the model-scale propeller used in the experimental campaign discussed in Chapter 2. The map data are comprised of the results from 100 simulations covering a range of ten propeller revolution rates from 1000 to 1900 RPM and ten advance ratios from 0.35 to 0.8. While the model propeller was a controllable pitch model, the pitch was held constant at the design pitch for the purpose of mapping, emulating a fixed-pitch propeller. The map generated using this procedure was further limited to a single ambient pressure (i.e. a single vessel draft) and quazi-steady operation in a single direction of travel. Figure 3.8 presents four maps based on the data set produced for the model propeller. Figure 3.8A shows a map of the torque coefficient predictions from panel method simulations against the propeller simulation inputs of advance ratio and revolution rate, and Figure 3.8B shows M-weighted noise predictions on the same pair of parameters. Figure 3.8C transforms the data from A and B such that noise is mapped over a torque coefficient and RPM parameter space. Figure 3.8D presents the map in its finalized form, relating shaft torque and RPM to cavitation source noise levels. Assuming an engine was connected to the propeller in a direct-drive configuration, the finalized map could be used alongside a emission map, such as the one shown in Figure 3.1, to simultaneously estimate chemical and noise emission levels for a given engine torque and speed combination.

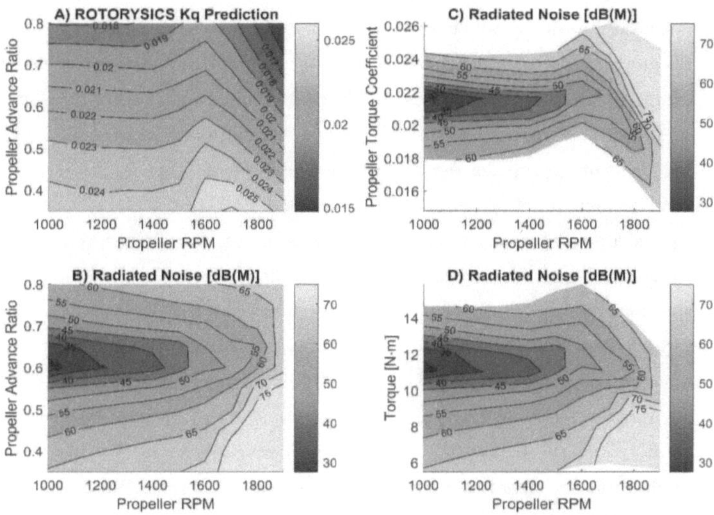

Figure 3.8: Cavitation noise mapping proof-of-concept showing (A) numerically-computed torque coefficients on an RPM-advance ratio parameter space, (B) M-weighted noise levels on an RPM-advance ratio parameter space, (C) M-weighted noise levels on an RPM-torque coefficient parameter space, and (D) the finalized noise map presenting the predicted noise levels on a dimensional RPM-torque parameter space.

The topology of the noise map generated with the present methodology is in agreement with the key findings of the scale model experiments. Experimental conditions C2, C4, and C5 saw the propeller operating at identical cavitation indices, and at the same revolution rates, with different advance ratios. At an advance ratio close to the designed condition (C2) the noise level was minimum, and tip vortex cavitation was observed in isolation. When the advance ratio was increased (C4) the propeller was under-loaded and a transition to a pressure side cavitation regime took place, resulting in increased noise. When the advance ratio was reduced (C5) the size of the tip vortex cavity increase and sheet cavitation appeared near the tip of the pressure side of the propeller blades, once again resulting in increased noise. The same trend is seen in the noise map Figure 3.8B: at a given revolution rate a local minimum in noise appears within the range of advance ratios. This optimal advance ratio is nearly independent of the propeller revolution rate, suggesting it is most likely a function of propeller design; however, the present map has not been rigorously validated with a sufficiently large experimental dataset, and it would be premature to comment on the significance of the numerical value of that optimal condition. Propeller torque was closely related to advance ratio in the simulation results, as shown in Figure 3.8A, and as a result the optimal torque also appears to be nearly independent of propeller revolution rate in Figure 3.8C and D.

The final proof-of-concept map presented in Figure 3.8D is successful in achieving equivalent functionality to chemical emission maps. A procedure combining panel method simulations with semi-empirical noise modelling was capable of uniquely relating shaft speed and torque pairs to cavitation noise estimates. Validation of the procedure remains forthcoming. Notably, the validation of any modelling procedure for generating practical noise emission maps should be done at full-scale, ideally with sea trial measurements.

Many vessels use controllable pitch propellers whose speed control strategy differs greatly from the fixed-pitch strategy considered herein. In the case that a controllable pitch propeller operates at a constant speed and vessel speed is controlled via the propeller's pitch, a map of noise like the one in the present work could likely be obtained over a toque-pitch space using nearly the same procedure. If the pitch and revolution rate are both varied, the noise map would need to be created over a three-parameter space including pitch, torque, and revolution rate.

3.7   Conclusions

We presented a methodology for generating a parametric map of cavitation noise for ships using pitch propellers on an RPM-torque parameter space. The mapping methodology relies on sweeping the parameter space with panel method simulations of the propeller, from which noise levels are estimated by semi-empirical noise models. We proposed a hybrid noise model combining Brown's formulation for the contribution of sheet-type cavities and the ETV-2 model for the contribution cavitating tip vortices to radiated sound.

We applied the modelling procedure developed for the mapping process to reproduce a set of scale model cavitating propeller experiments. The panel method was successful in predicting the cavitation regimes seen in the experiments, but did not accurately reproduce the patterns. We applied three different semi-empirical noise models to predict the emitted noise spectra based on the simulations and compared the results to experimental measurements: Brown's formula, the ETV-2 model, and our hybrid formulation intended to account for multiple cavitation regimes. A hybrid formulation was necessary to cover a wide enough range of operating conditions to form an effective map.

A proof-of-concept for the present methodology was executed based on a model-scale propeller in uniform inflow. Simulations used for the mapping procedure were compared to experimental measurements of the propeller both in terms of cavitation patterns and acoustic signals. The proof-of-concept successfully related each shaft torque and speed pair to a single estimated cavitation noise level. The topology of the generated map suggests that an optimal torque exists for the model propeller used in the present study, and that the optimum is nearly independent of the propeller's revolution rate. This result is not surprising, given that most propellers are designed to perform efficiently at under a specific set of operating conditions.

A full-scale validation, ideally with sea trial data, is necessary to bring the present strategy from a proof-of-concept to a validated procedure.

Chapter 4: Influence of Propellers and Operating Conditions on Underwater Radiated Noise from Coastal Ferry Vessels

Duncan McIntyre[1], Waltfred Lee[1], Héloïse Frouin-Mouy[2], Dave Hannay[2], Peter Oshkai[1]

[1] Department of Mechanical Engineering, University of Victoria, Canada

[2] JASCO Applied Sciences Ltd., Victoria, Canada

4.2 Keywords

Underwater radiated noise; cavitation-induced noise; field measurement

4.3 Introduction

As the world population grows, so too does marine vessel traffic in the world's oceans. Ships represent the largest source of anthropogenic underwater noise [17], and growing shipping activity threatens marine ecosystems where fauna favour sound for sensing and communication [6], [8], [10]. In the short-term auditory masking from these acoustic disturbances can result in disruption of breeding in animals that use sound during mating and reproduction, and disruption of foraging in animals that use sound to detect prey, resulting in lower survival rates of at-risk marine species when exposure is chronic. Historically, interest in underwater radiated noise from ships was primarily military; however, an increase in academic, regulatory, and commercial commitment to mitigating underwater noise pollution has resulted in a greatly expanded civilian-focused body of research in the past decade. Most works to date have focused on source modelling [30], [39], [40], [68], [77], environmental assessment [9], or regulatory-level management strategies [79], [80]. While vessel slow-downs have been shown to reduce total ambient noise in critical habitat areas [79], vessel speed is not universally correlated with increased radiated noise. More work is required to enable the development of mitigation strategies for radiated noise from particular vessels. While source modelling provides the potential for noise mitigation strategies of this type, several factors have so far prevented their adoption by fleet operators. Implementation of numerical noise modelling strategies is inhibited by reliability questions, computational expense, and the inability for fleet operators to obtain detailed propeller geometry information from manufacturers. Sea trials relating noise to vessel operation, on the other hand, are infeasible for most fleet operators. As a result, it remains challenging to relate underwater radiated noise to specific vessel operating conditions, and it therefore also remains challenging to develop mitigation strategies at the level of individual vessels.

Underwater radiated noise from ships is primarily generated by machinery, turbulence, and cavitation. Machinery noise is generated by a vessel's on-board mechanical systems, most notably the drive train. Vessels with fixed-pitch propellers rely on their drive train to control their speed through water, and the machinery-induced component of radiated noise from these vessels depends on their speed. Vessels with controllable-pitch propellers, on the other hand, use the pitch of their propellers for speed control; they typically operate their propellers and engines at constant revolution rates, and machinery noise from these vessels is therefore expected to be speed independent. Turbulence noise is generated around both the propeller and the hull, and therefore depends on both propeller operating conditions and vessel speed through water. As a mechanism of sound generation, turbulence tends to be much quieter than either machinery or cavitation. Cavitation noise depends on similar conditions to turbulence noise with the addition of vessel draft, since cavitation inception depends on the local static pressure around the propeller. Because cavitation commonly occurs during typical propeller operation and produces sound of a much greater amplitude than turbulence, it is often the dominant source of noise emitted by marine vessels [18]. Operating under the assumption that cavitation noise dominated the radiated noise signature from all participating vessels, those operating conditions known to be associated with cavitation noise were the focus of the present work. Common dimensionless parameters characterizing the operating conditions associated with cavitation noise are outlined in Table 4.1.

Table 4.1: Typical dimensionless quantities used in the analysis of URN from marine propeller cavitation

Name	Symbol	Definition	Significance
Nominal cavitation index	σ_n	$\dfrac{p_{ref} - p_{vapour}}{0.5\rho n^2 D^2}$	The cavitation index of a propeller is a representation of the likelihood for cavitation. It accounts for a combination of the ambient pressure at the propeller depth p_{ref} relative to the vapour pressure of seawater p_{vapour} and the tip speed of the propeller. Since the ambient pressure takes the local hydrostatic pressure into account, it accounts for propeller rotation speed n and diameter D, as well as vessel draft. Density ρ is generally assumed to be constant.
Wake fraction	w	$1 - \dfrac{V_A}{V}$	The wake fraction relates the average advance velocity V_A of the propeller through the water to the speed V of the vessel through the water. It depends mainly on the shape of the ship's hull.
Advance ratio	J	$\dfrac{V_A}{nD}$	The advance ratio represents the distance a propeller travels through the water each revolution and is normalized on the propeller diameter. For variable pitch propellers that always run at the same rotation speed it is functionally a dimensionless advance velocity.
Dimensionless pitch	P/D	$\dfrac{P}{D}$	The distance the propeller would travel through water in a single revolution if no slip occurred. It is conventionally non-dimensionalized on the propeller's diameter. Variable pitch propellers are geometrically optimized in terms of twist and skew for their design pitch, so a change in pitch results in off-design operation.
Slip ratio	S	$1 - \dfrac{J}{P/D}$	The slip ratio represents the difference between the true advance rate and the ideal advance rate defined by the pitch of the propeller.

This work aims to shed light on the relationships between vessel operating conditions and underwater radiated noise using a comprehensive set of far-field acoustic measurements of passenger ferries operating off the west coast of Canada. The vessels operated commercially while passing by a hydrophone array along a predetermined path, facilitating systematic measurements. Captains operated the vessels at a range of speed settings and recorded detailed vessel operating data, which was combined with acoustic measurements.

4.4　Materials and methods

Far-field underwater acoustic measurements of eight coastal ferry vessels were collected from vessels operating at varied speed settings while passing through a constricted channel. Spectral data were acquired with a JASCO Applied Sciences Autonomous Multichannel Acoustic Recorder (AMAR G3). Vessels travelled a set track passing whose closest point of approach was 160 m from the recorder. Adherence to the track was confirmed for each measurement using geographic positioning data broadcast by each vessel over the Automatic Identification System (AIS). Measurements were analysed for vessel positions with a $\pm 30°$ azimuth angle centred on the recorder at frequencies ranging from 10 Hz to 31.5 kHz. Power spectral densities (PSDs) at the receiver location were computing by sliding one-second Fast Fourier Transforms (FFTs) with power-normalizing Hanning windows with a 50% overlap. Background noise was corrected for using averaged noise levels from a pair of one-minute windows, the first one minute prior to the vessel entering the entrance tunnel to the channel, and the second one minute after the vessel left the exit zone. Measurements with interference from other vessels, high levels of background noise, or where the vessel deviated from the prescribed measurement procedure were rejected either automatically based on AIS data or manually during a quality control review.

The spectra analysed throughout the present work are given in terms of Radiated Noise Levels (RNL) and have been back-propagated to a point source located at the GPS reported vessel position assuming spherical spreading loss in accordance with ANSI standard s12.64-2009. The present data have therefore not been corrected for sea surface or seabed reflections. Where possible, vessels were operated at five different speed settings that are defined qualitatively. Speeds categorized as Full Away (FA) represent the upper end of practical service speeds, Service Speed (SS) is the typical speed for the route, 2 kn Reduction (r2) and 4 kn Reduction (r4) are 2 kn and 4 kn slower than Service Speed respectively, and Half Speed (HS) is half of Service speed. While each speed setting represents a desired speed and corresponds to a particular vessel operation setup, there is variability within in each setting in terms of both actual vessel speed and other vessel operating conditions.

Table 4.2 provides a description of the eight vessels for which data were collected. Five of the eight vessels were fitted with controllable pitch (CP) propellers which were operated at constant revolution rates and used pitch manipulation for speed control. The remaining three vessels were fitted with fixed pitch (FP) propellers, and relied on revolution rate manipulation for speed control. Vessel 1 was operated with only the aft (depending on the travel direction) propeller while travelling through the measurement area; all other vessels were operated with all propellers powered. Vessels 5 and 6, as well as 7 and 8, were sister ships with the same design. Vessel 7 was operated with only three of its four main engines during the test period, but operated both propellers and ran at speeds comparable to its sister ship. Previous analysis has found significant positive correlation between speed and broadband underwater radiated noise for Vessels 2, 4, and 5, while significant anti-correlation between speed and noise was found for Vessels 1 and 8 [81].

Not all information required to assess cavitation noise for each vessel was available to the operator. Detailed propeller geometry was only available for vessel 1, and that vessel is therefore the primary focus of the present work. It was also necessary to estimate the speed of water incoming into the propeller; however, limited information was available to calculate the wake fraction. Therefore the simple formula of Taylor, as given in [70], is adopted in the present work:

$$w = 0.5C_B - 0.05, \tag{4.1}$$

where C_B is the block coefficient of the vessel (the volume over the product of the length, beam, and draft). As Carleton notes in [70], this formula is simple and convenient, but due to advances in hull design since the 1930s it has limited accuracy and should not be considered a reliable tool for design. For the purposes of a regression or correlation analysis, however, the accuracy of the wake fraction estimate is unlikely to be critical.

In addition to the more commonly used dimensionless quantities, a version of the draft normalized on the propeller diameter (T/D) is used in the present analysis for consistency with the pitch and advance ratio. As is conventional, all pitches are specified at 70% of the propeller radius $(0.7R)$ unless otherwise noted.

Table 4.2: Specifications of the eight vessels that took part in the study as well as the number of measurements of each that were available for the analysis

Vessel	Year Built	Length	Direction	Propellers	Measurements
1	2008	160 m	Double	1 Fore + 1 Aft, CP $D = 5.0$ m, 4 bladed	44 accepted 33 rejected
2	1965	85 m	Single	2 Fore + 2 Aft, FP $D = 1.68$ m, 4 bladed	50 accepted 23 rejected
3	1992	96 m	Double	2 Fore + 2 Aft, CP $D = 2.0$ m, 4 bladed	28 accepted 67 rejected
4	1964	129.9 m	Single	2 Aft, CP $D = 2.9$ m, 4 bladed	68 accepted 42 rejected
5	2016	107 m	Double	2 Fore + 2 Aft, FP $D = 2.45$ m, 4 bladed	17 accepted 24 rejected
6	2017	107 m	Double	2 Fore + 2 Aft, FP $D = 2.45$ m, 4 bladed	9 accepted 8 rejected
7	1993	167.5 m	Single	2 Aft, CP $D = 3.4$ m, 4 bladed	53 accepted 31 rejected
8	1994	167.5 m	Single	2 Aft, CP ⌀ 3.4 m, 4 bladed	49 accepted 30 rejected

Experiments have shown that different cavitation regimes result from propellers under different loading conditions, and that these unique regimes have different acoustic signatures and relative loudness [63]. To account for the different cavitation regimes that occur when the propeller is differently loaded, a pair of modified slip quantities referred to hereafter as "overslip" S_O and "underslip" S_U were defined. The former quantifies how much more slip the propeller is experiencing that it would under designed conditions, while the latter quantifies how much less slip it experiences:

$$S_O = \begin{cases} 0, & S < S_{\text{design}} \\ S - S_{\text{design}}, & S \geq S_{\text{design}} \end{cases} \quad S_U = \begin{cases} S_{\text{design}} - S, & S < S_{\text{design}} \\ 0, & S \geq S_{\text{design}} \end{cases}. \quad (4.2)$$

To predict the existence and extent of propeller-induced cavitation requires more information than is presented in Table 4.1, including the complete blade geometry and wake shape. In many cases it may not be possible to obtain this information; in fact, even the information listed may not be available even to the vessel operator. The wake fraction depends on the hull form, which may be estimated from the geometry of the vessel's hull (via the stability booklet information), but not is generally known. The pitch is a part of the propeller design which is generally proprietary information, and is in some cases not available to the owner of the vessel, as is the case for some of the vessels considered in the present work. As such, an attempt has been made to construct models without the need for detailed propeller and wake information. Among the vessels considered in the present work, detailed propeller information was only available for Vessel 1.

4.5 Analysis

4.5.1 Narrow-band RNL trends with vessel speed

As a first step in understanding the URN behaviour of the eight vessels, broadband levels were examined on a vessel-by-vessel basis. In order to protect the privacy of the vessel operator, absolute levels have been normalized on an arbitrary reference. The relationship between vessel speed and radiated noise was investigated across the recorded frequency range. Surface plots of RNL across vessel operating speeds and frequencies are presented in Figure 4.1. Vessels 2, 4, 5, and 6 all showed positive correlations between vessel speed and RNL for most frequencies, while vessels 1, 3, 7, and 8 showed anti-correlations for most frequencies; however, these trends did not hold over the entire frequency range. The factors determining the broadband RNL behaviour of a given vessel were not clear from the available data. Neither observed relationship was associated with a particular propeller layout, vessel size, or vessel age. Vessels 1, 3, 7, and 8, which all showed anti-correlation between speed and noise, used controllable pitch propellers; however, another vessel with controllable pitch propellers, vessel 4, displayed a positive speed-RNL correlation. This finding is particularly important from a regulatory perspective: while slowdowns have been found to result in net reductions in ambient noise levels from marine vessels [82], [83], it is clear that not all vessels benefit from speed reduction.

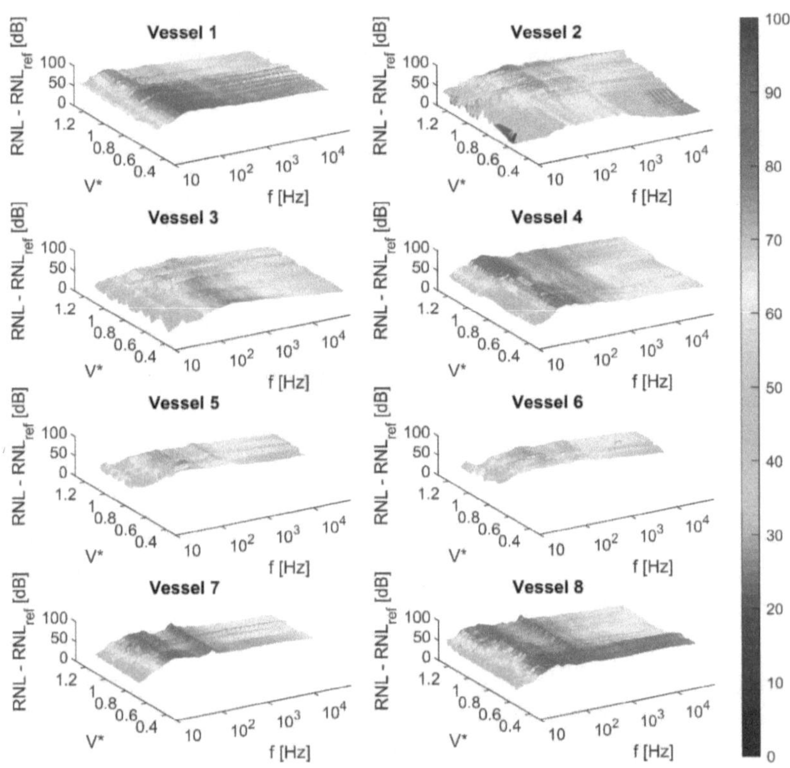

Figure 4.1: Surface plots showing the relationship between vessel speed, frequency, and radiated noise level. Velocities are shown normalized on each vessel's average service speed. RNL has been normalized on an arbitrary RNL value to protect the intellectual property rights of the fleet owner. In addition to the levels of noise depending on the speed of a vessel, the distribution of acoustic energy through the spectrum varies with vessel speed.

The distribution of acoustic energy between frequencies was also observed to be vary between operating conditions. To better visualise this, a selection of single-frequency-band cross sections of the surfaces presented in Figure 4.1 are shown in Figure 4.2. Vessels 1 and 8 show particularly strong spectral shape changes between high- and low-speed operation, which is explored in greater detail in section 4.5.3. Vessel 7 was notably a sister ship to Vessel 8, and would be expected to exhibit similar characteristics in terms of radiated noise had a larger range of speeds been captured for measurement. Low-frequency noise tended to follow a different speed trend compared to mid- and high-frequency noise for all eight vessels, suggesting a difference in the source; this result is expected where cavitation noise is significant and dominates the mid- and high-frequency ranges while mechanical noise dominates the low-frequency domain. A notable exception is tonal noise at the blade passing frequency (the frequency with which the propeller blades pass through the wake), which is hydroacousctic in nature and caused by fluctuations in flow as the propeller blades pass through the wake.

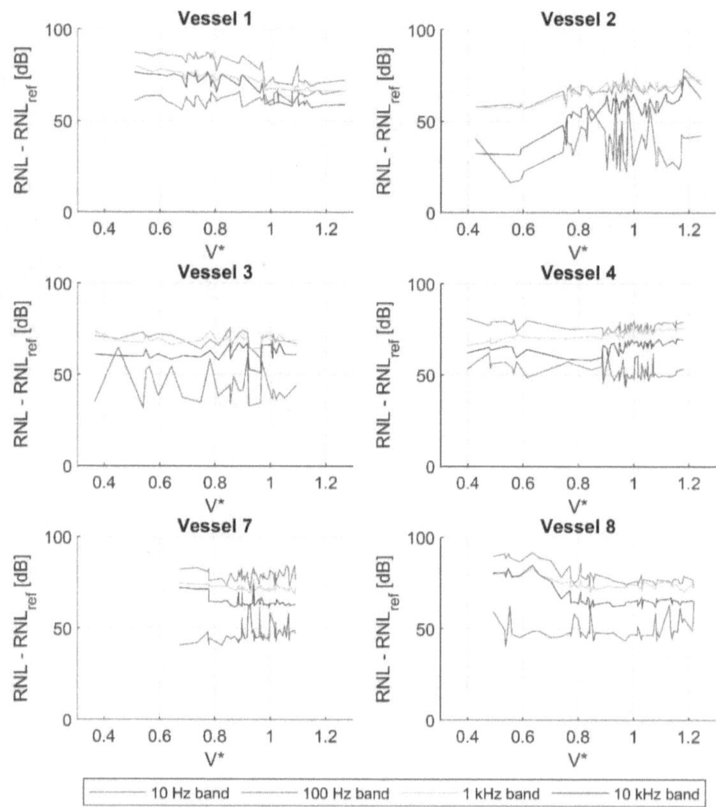

Figure 4.2: Third octave band radiated noise levels at varied vessel speeds for six different vessels. Vessels 1, 2, and 3 have controllable pitch propellers, while vessel 4 has fixed pitch propellers. Vessels 2 and 3 belong to the same sub-class but operated with different engine configurations. Speeds are given normalised on the given vessel's service speed. Vessels 5 and 6 have been excluded due to the relatively small range of measured operating conditions available for those vessels.

4.5.2 Multivariable linear regression analysis

Linear regression analysis was used in order to explore the influence of various vessel operating parameters on radiated noise. Due to the limited information available on vessel design, the analysis was limited to vessel 1. In order to mitigate error due to potential mismatches between recorded and real operating conditions, measurements that deviated by more than 50% from the travel speed or propeller pitch prescribed by their listed speed setup were excluded from this and all further analysis in following sections. This reduced the total number of available measurements for the regression analysis of vessel 1 to 27. To select the parameters used in the regression model of RNL, the correlation coefficients of dimensionless operating parameters and each third-octave band RNL were examined, as shown in Figure 4.3. Low frequency noise, especially at 20 Hz and below, was found to be minimally correlated with vessel operation, suggesting that it is dominated by mechanical noise. The 10 Hz band include the blade passing frequency of 9.3 Hz for this vessel, and is therefore expected to be dominated by noise from blade passage effects; at this frequency RNL was, uniquely, positively correlated with speed. Above 20 Hz the cavitation index was found to be strongly correlated with RNL, while the advance ratio (equivalent in this case to speed) was found to be strongly anti-correlated. The anti-correlation with advance ratio was expected for this particular vessel. Since the vessel operated with a constant propeller revolution rate, the cavitation number was determined almost entirely by the vessel draft, and it was therefore excluded from the regression analysis. The observed anti-correlation between draft and noise is consistent with propeller-induced cavitation noise. Given that the slip terms were not strongly correlated with RNL and that slip is a linear combination of advance ratio and pitch, slip was excluded from the regression analysis.

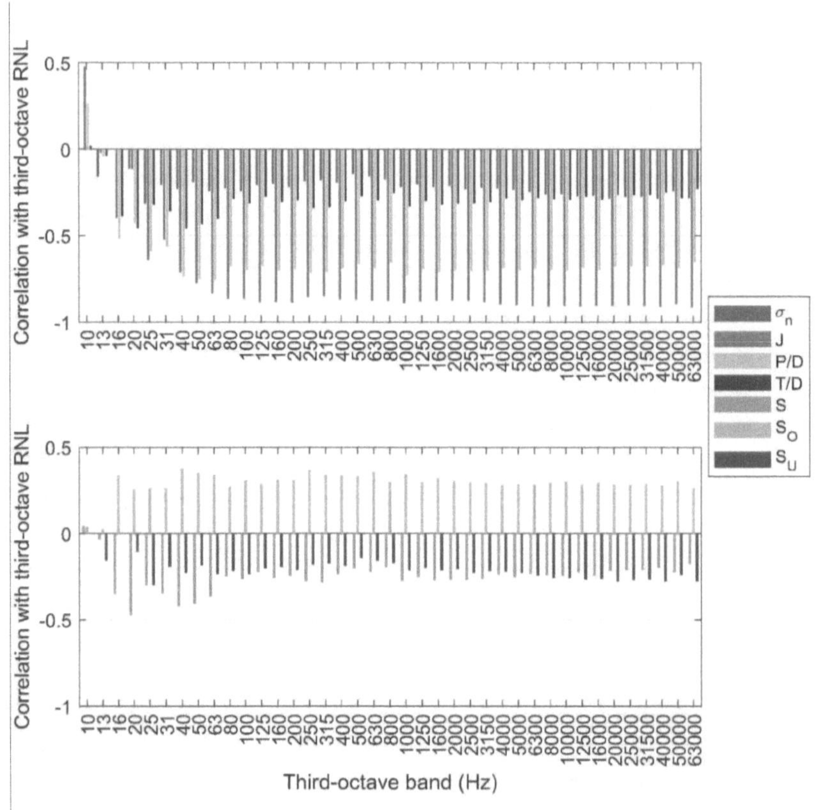

Figure 4.3: Correlations between ship operation parameters and third-octave band radiated noise levels for vessel 1. The selected parameters are all relevant to the production of cavitation but are not independent from each other._

A multivariable linear regression was performed for RNL at each third-octave band using a total of 27 measurements, resulting in 39 regression models of the form:

$$Third\text{-}Octave\ RNL_f = C_0 + C_{1,f} + C_{2,f}J + C_{3,f}(P/D) + C_{4,f}(T/D)\ [\text{dB}] \qquad (4.3)$$

The constant value of C_0 is was chosen arbitrarily to allow the C_1 term in each equation to be more comparable in magnitude to the other terms for plotting. The resulting model coefficients are plotted against frequency in Figure 4.4a, while the R^2 values of each regression are plotted in Figure 4.5. Comparing the magnitudes of the regression model coefficients gave some indication as to their influence, although care should be taken to note that the range of values for each of the regression terms is not the same. A comparison of the values of each term in the linear regression using the mean value of each variable from all 27 measurements is given in Figure 4.4b to present a fairer comparison of the influence of each term in the model on the output. From this comparison it can be seen that the influence of pitch (C_3) and draft (C_4) was similar across most frequencies above 30 Hz, while the constant and advance ratio terms in the model varied more substantially with frequency.

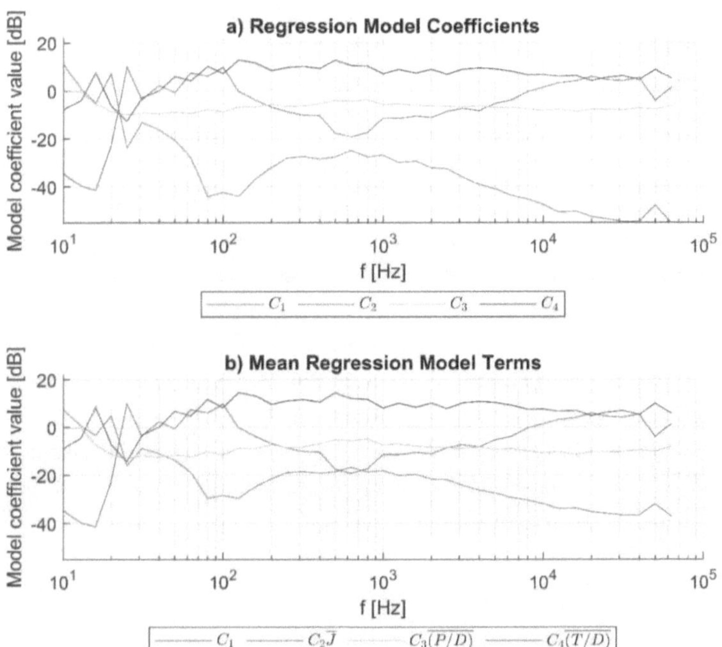

Figure 4.4: a) Coefficients of the single frequency band regression models for third-octave band RNL as a function of advance ratio, pitch, and draft of vessel 1. Each coefficient has units of decibels. b) The products of each model coefficient with the mean value of their corresponding variables from the 27 measurements used to generate the regression. The resulting values give a good indication of the relative influence of each term on the output of the regression for a given frequency.

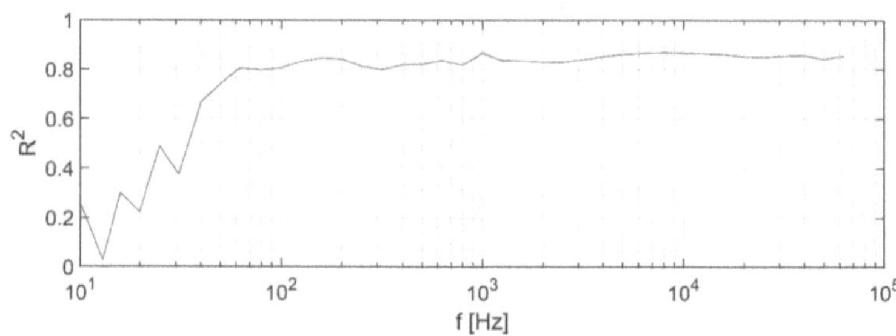

Figure 4.5: R^2 values of the single frequency band regression models for vessel 1 presented as a function of frequency.

It is apparent from Figure 4.5 that a regression of the form presented in Eqn. (4.3) is not suitable for frequencies below 100 Hz. Above this frequency, the trend of each coefficient as shown in Figure 4.4a is characterized by a pair of inflections. In the absence of a reliable physical explanation for these trends, it was decided that the regressions for individual frequency bands should be combined by fitting a cubic polynomial to each of the regression coefficient trend lines. The resultant model is a one equation dedicated model that can predict an RNL spectrum from vessel 1 given a set of operating conditions as inputs.

Comparisons between five experimentally measured spectra and spectra produced by this regression formula are shown in Figure 4.6. The five measurements were chosen as examples because they were representative of five handle settings used in the operation of the vessel, not because of the relative goodness or badness of the regression fits. The detailed regression model successfully predicts both the general shape of the spectrum and its magnitude for most measurements.

To check the accuracy of the regression in reproducing the data, the mean, maximum absolute, and RMS error in predicted RNL over the applicable range of frequencies was computed for each of the 27 measurements used to generate the regression formula. The results are shown in Figure 4.7, where the errors are plotted against the speed through water of the individual measurements. The spectrally averaged error is, with a single exception, within ± 5 dB, while some individual frequency bands are over or underestimated by up to 10 dB. The lack of a clear trend within the error suggests that the regression formula is equally applicable across the entire speed range.

The regression model that has been generated using this procedure is unlikely to be generalizable, since it always predicts an anti-correlation relationship between ship speed and radiated noise. The model is heavily reliant of the constant term in the equation, which has no physical significance; it therefore provides little insight into the source of the noise.

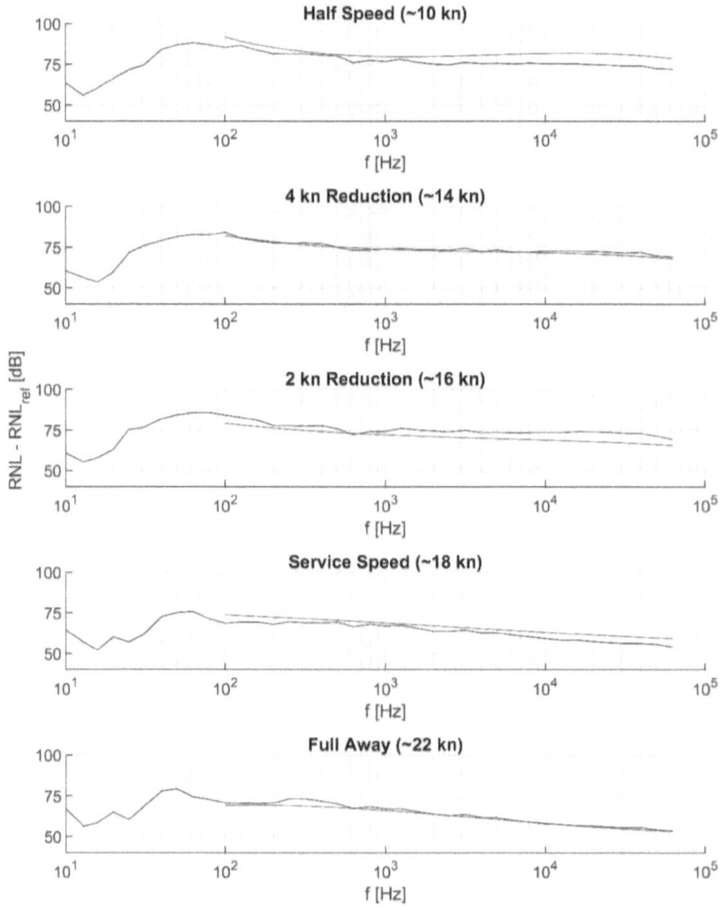

Figure 4.6: Comparison between measured (blue) and detailed regression model predicted (red) third octave band RNL spectra for the five measurements whose speed through water was closest to the average for their individual handle settings (Half Speed, 4 kn Reduction, 2 kn Reduction, Service Speed, and Full Away) of vessel 1.

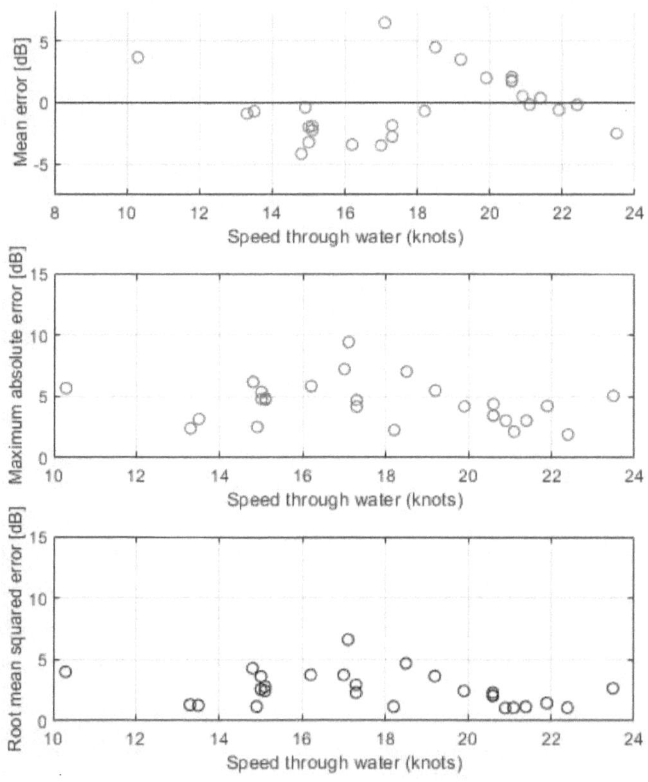

Figure 4.7: Regression spectral mean, maximum absolute, and mean squared errors for each of the 27 measurements used in the generation of the regression for Vessel 1.

4.5.3 Radiated noise regime analysis based on broadband spectral features

The relationships between speed and RNL shown in Figure 4.1 and Figure 4.2 show evidence of a regime change in acoustic emission between high- and low-speed operation for Vessels 1 and 8. For vessel 1 a common set of features determined to belong to a high-speed regime (HSR) and a low-speed regime (LSR) were observed in all spectra collected from operating speeds above 110% of service speed and below 80% of service speed respectively ($V^* > 1.1, V^* < 0.8$). The corresponding thresholds for vessel 8 were found to be 90% of service speed and 60% of service speed. These four sets of measurements were used to define the set of spectral features characterizing each regime. Figure 4.8 and Figure 4.9 show the HSR and LSR defining measurements for Vessels 1 and 8. An averaged spectrum is also shown for each set of data in order to reduce noise and make visualization of the spectral features easier. The salient features of these regimes are described visually in Figure 4.10 and Figure 4.11. As with the regression analysis, measurements that deviated by more than 50% from the travel speed or propeller pitch prescribed by their listed speed setup were excluded from the analysis; however, given the small number of available measurements in the low-speed regimes, the speed ranges were checked against the measurements that were excluded. Even with the larger data set, which included six additional members of the low-speed regime group in the case of Vessel 1, all measurements belonging to one of the defining speed ranges shared these salient features.

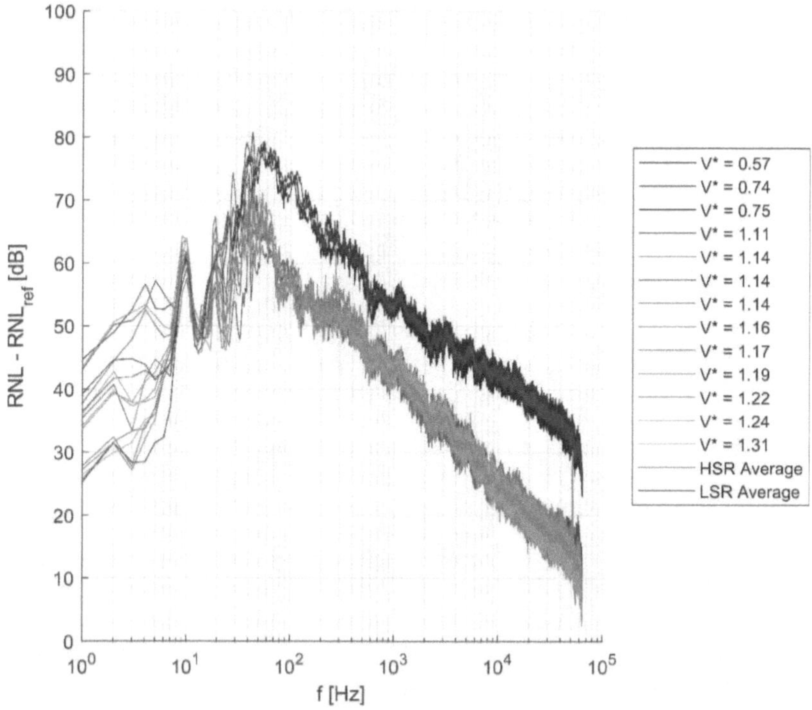

Figure 4.8: PSDs of RNL from Vessel 1 for vessel speeds characterizing the high- and low-speed noise emission regimes. The bright coloured lines are the averages of these spectra, which maintain the salient features of these spectra while reducing noise.

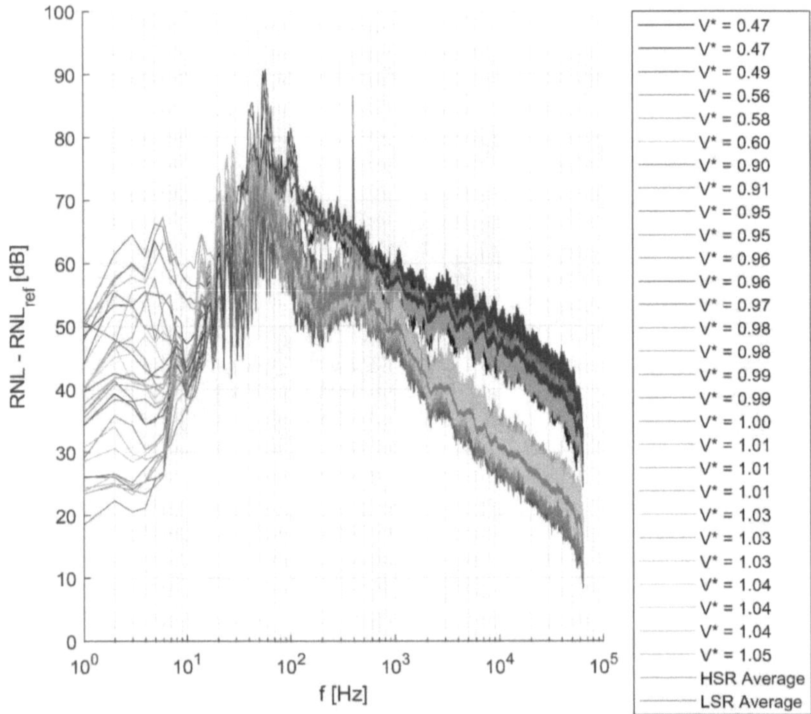

Figure 4.9: PSDs of RNL from Vessel 8 for vessel speeds characterizing the high- and low-speed noise emission regimes (HSR and LSR respectively). The bright coloured lines are the averages of these spectra, which maintain the salient features of these spectra while reducing noise.

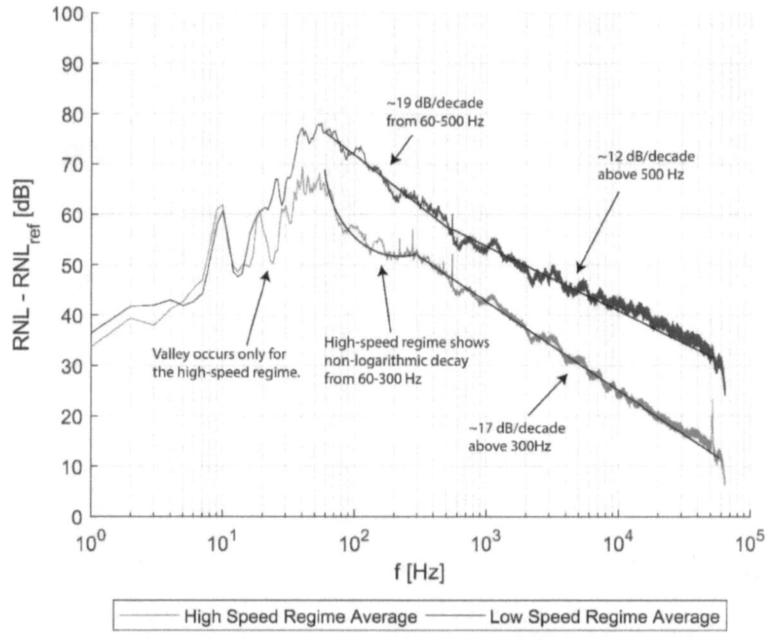

Figure 4.10: Distinguishing features of the acoustic signatures of the high- and low-speed operating regimes of Vessel 1. These features were used to sort operating conditions into one of the two regimes. When examined individually, these features could be clearly seen in each of the acoustic signatures that were averaged to get the representative signatures used for identification.

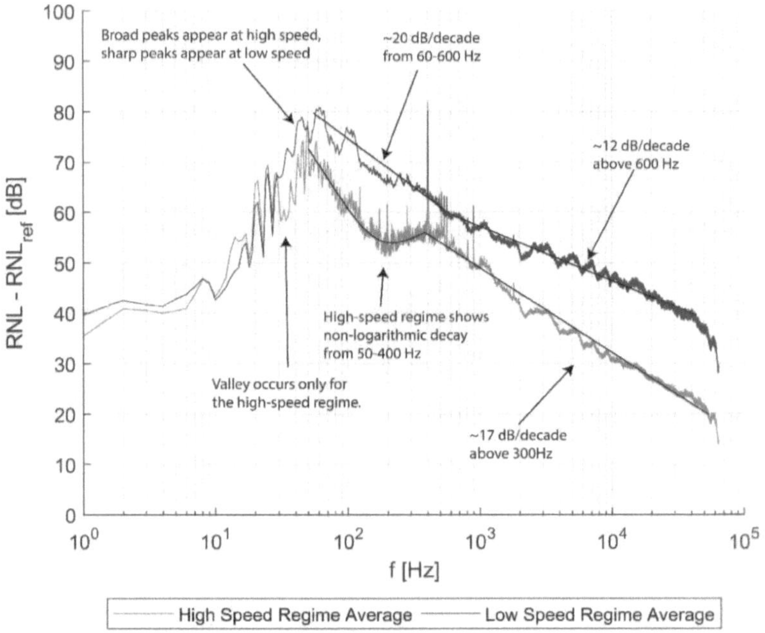

Figure 4.11: Distinguishing features of the acoustic signatures of the high- and low-speed operating regimes of Vessel 8. These features were used to sort operating conditions into one of the two regimes. When examined individually, these features could be clearly seen in each of the acoustic signatures that were averaged to get the representative signatures used for identification.

The high-speed regimes of each vessel share several spectral features in common in the high- and mid-frequency ranges, including the slope of the logarithmic decay trend above 300 Hz, a non-logarithmic trend from the high tens to low hundreds of hertz, and a valley not seen at low speeds in a frequency range just below the peak frequency. Similarly, the low-speed regimes of the two vessels both show a pair of logarithmic decay trends with a slope reduction around 600 Hz. The low frequency noise signatures of the two vessels, in contrast, have little in common. Given that propeller noise, especially cavitation noise, tends to dominate the mid- and high-frequency RNL spectra of ships, the most likely cause of the similarity in the behaviour of the two vessels is that each of the two acoustic regimes corresponds to a particular cavitation regime. The spectra of the high-speed regime are similar to the tip vortex cavitation emission spectra observed in scaled experiments of a controllable pitch propeller discussed in [63], while the low-speed regime spectra resemble the pressure side leading edge sheet cavitation observed for reduced-pitch operation in those experiments.

Among the measurements that did not belong to one of the defining speed sets, some were identifiable as members of one of the two regimes, as is the case for the measurements shown in Figure 4.12, which was classified as a belonging to the low-speed regime, and Figure 4.13, which was classified as belonging to the high-speed regime. Other measurements exhibited a mixture of characteristics from both regimes. It was unclear if these measurements were the result of time-averaging of signals for which the ships propellers underwent a transition between the two acoustic emission regimes, or if they represented another regime or set of regimes entirely. For the purposes of this study, all acoustic measurements not categorized as belonging to the high- or low-speed regimes were considered to belong to a transitional regime. One example of such a measurement is shown in Figure 4.14.

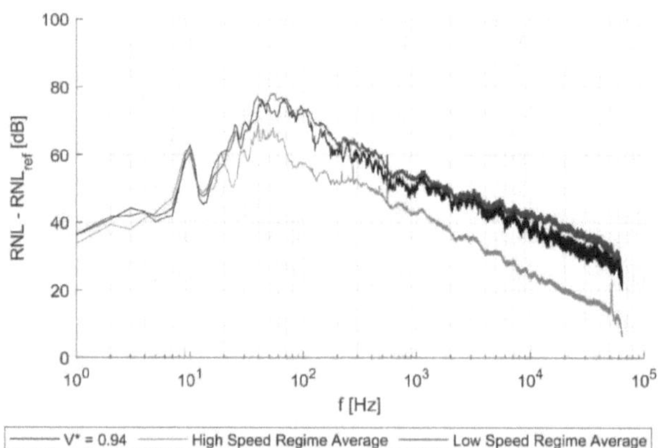

Figure 4.12: An example of an operating condition outside the defining speed ranges (94% of service speed) under which Vessel 1 appeared to exhibit the characteristic acoustic signature of its low-speed operating regime. Although the absolute noise level is lower than average for the low-speed regime, the salient features of the acoustic signature can be clearly seen.

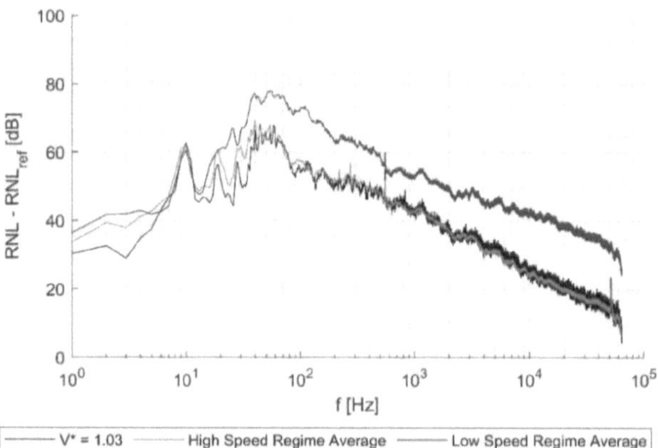

Figure 4.13: An example of an operating condition outside the defining speed ranges (103% of service speed) under which Vessel 1 appeared to exhibit the characteristic acoustic signature of its low-speed operating regime. The salient features of the acoustic signature can be clearly seen.

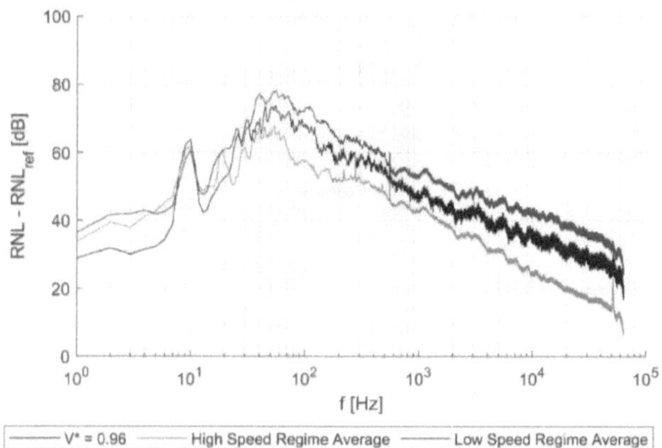

Figure 4.14: An example of an operating condition (96% of service speed) under which Vessel 1 radiated an acoustic signature that was characterised by a mix of features from both the high- and low-speed regimes. Operating conditions that exhibited spectral features of both regimes were categorized as transitional. The present example deviated notably from either regime in the low frequency range between 10 and 50 Hz and exhibited decay trends between those characterizing each regime.

To asses the relationship between the acoustic (and presumed cavitation) regimes and operating conditions other than speed through water, the distribution of acoustic measurements was assessed against draft, pitch, and cavitation index. The results for vessel 1 are shown graphically in Figure 4.15. The spectral features associated with the low-speed regime align well with those of the pressure side cavitation regime found experimentally in [63] to be associated with reduced pitch propeller operation; the observed connection between pitch and acoustic regime changes is strong evidence that the low-speed acoustic regime was associated with pressure side propeller cavitation. As expected for a transition from a combination of tip vortex and suction side cavitation to pressure side cavitation, there was no indication that the cavitation regime was related to either the draft or the cavitation index.

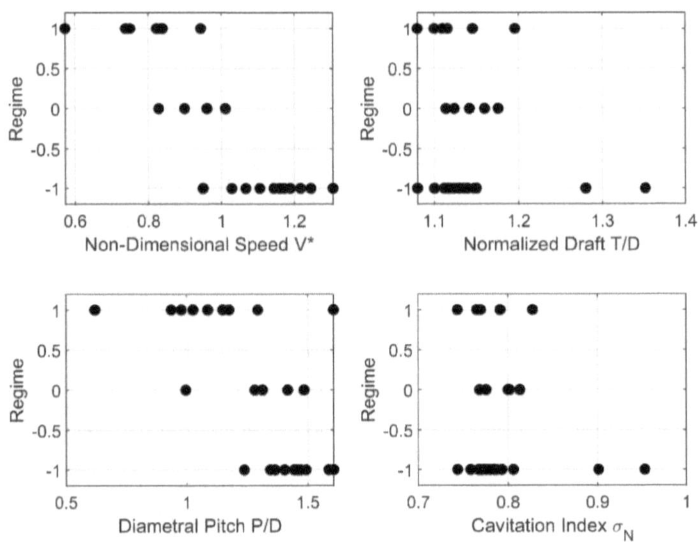

Figure 4.15: Non-dimensional speed (top left), normalized draft (top right), diametral pitch (bottom left) and cavitation index (bottom right) of each measured operating condition shown sorted into three acoustic-regime bins. No single parameter or combination of parameters was sufficient to determine which acoustic regime an operating condition was expected to fall within.

4.5.4 Narrow-band spectral features

Low-frequency spectral features revealed distinct regimes associated with speed ranges, but little insight into the physical mechanisms resulting in those regimes. To investigate the regimes further, de-trended spectra were examined for the low- and high-speed regimes of vessels 1 and 8 defined in 0. Vessels 1 and 8 both employed controllable pitch propellers, and their engine and propeller revolution rates were maintained constant across all speed ranges; by selecting these vessels for the present analysis engine noise was eliminated as a source candidate for spectral features that change with speed. De-trending was accomplished by subtracting the baseline as shown in Figure 4.16, which were computed applying shape-preserving piecewise cubic interpolation and quadratic fit smoothing.

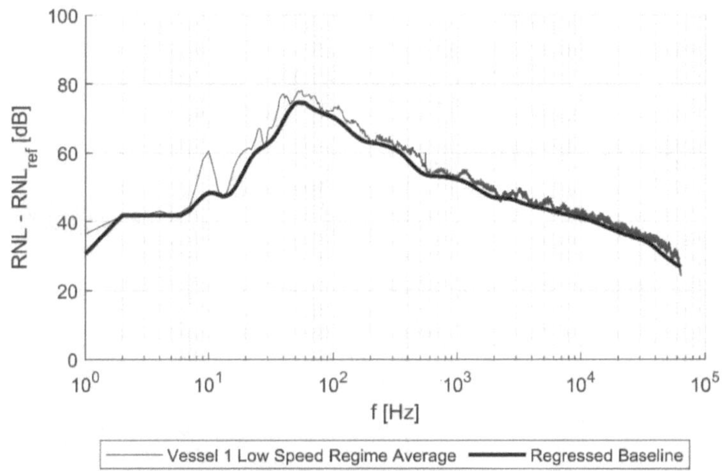

Figure 4.16: Averaged RNL spectrum of measurements classified as belonging to the low-speed regime of vessel 8 and the baseline used for de-trending the low-speed regime

The resulting de-trended noise spectra show narrow-band features of the acoustic regimes in isolation, allowing comparison between speeds and vessels. The absolute magnitudes of the peaks were not considered because each plot was de-trended using a different baseline; however, the location, width, and size relative to adjacent peaks is of interest. Figure 4.17 presents the de-trended noise spectra of vessel 1 and 8 at low and high-speed regimes. Narrow-band spectral features at frequencies above 550 Hz appear to be independent of vessel type and speed. This suggests that the narrow-band peaks in the high-frequency range come from sources common to all measurements; possible sources include common cavitation structures, background noise, or artifacts from filtering and correction. While broad spectral features were found to show discrete changes indicative of acoustic regimes between vessel speeds, narrow-band features at frequencies above approximately 550 Hz were observed to be largely unaffected by vessel operation, indicating that frequency-binned data are sufficient for analysing the radiated noise regimes of these vessels when the mid- and high-frequency ranges are of primary interest.

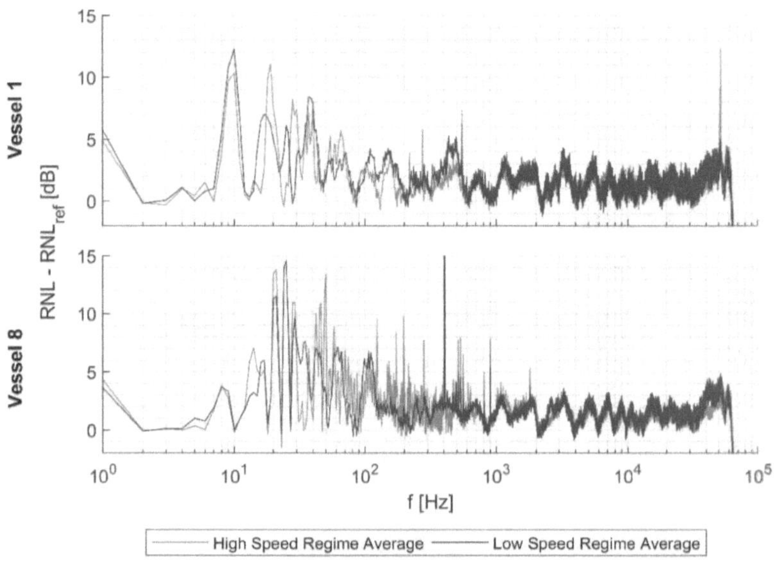

Figure 4.17: Full de-trended noise spectra of (a) Vessel 1 and (b) Vessel 8 at low and high-speed regimes. Features at frequencies above 550 Hz show little variation between the two different speed regimes for each vessel, while some of the lower-frequency features change in shape or appear and disappear between the speed regimes.

Notable changes in the narrow-band spectral features between vessels, and between speed regimes for the same vessel, were observed for frequencies below 550 Hz. In order to present a better view of the narrow-band spectral features at low frequencies the de-trended noise spectra is replotted with a linearly scaled x-axis of up to 300 Hz in Figure 4.18. Peaks with similar widths and locations are likely to be from speed-invariant sources, the majority of which are expected to be mechanical noise associated with engines. The blade passing frequencies of vessels 1 and 8 were 9.3 Hz and 13.7 Hz respectively, and peaks at these frequencies and their first multiples were assumed in this analysis to be associated with the blade passage phenomenon. Individual narrow-band features that change between speed regimes are identified with alphanumeric designations. These changing features are likely influenced by changing cavitation and flow features. Several narrow peaks between 100 and 300 Hz, labeled 1c, 1d, 8b, 8c, 8d, and 8e in Figure 4.18, were observed exclusively for the high-speed regimes of both vessels, and may be the result of a specific cavitation regime. The tonal nature of these spectral features suggests tip vortex cavitation as the most likely source, as other cavitation regimes produce primarily broadband noise [50].

In addition to features that only existed in the RNL spectra of one of the two regimes, some features changed in nature between regimes. The features labeled 1a and 8a show a change in nature between the low- and high-speed noise regimes; 1a existed as peak at high-speed and a valley at low-speed operation, while 8a behaved in the opposite manner. The feature labeled 1b also changed behaviour from a single broad hump at low-speed to a collection of 3 humps at high-speed. The source of these spectral features, as well as the cause of their change in behavior between regimes, is unclear.

The de-trended RNL spectra from vessel 8 shown in Figure 4.18 is modulated by a frequency of approximately 4 Hz. This modulation is not an artifact of the de-trending process; it can be seen in the averaged spectra shown in Figure 4.11 and the individual measurement spectra plotted in Figure 4.9. The same modulation also appears in spectra from measurements of high-speed operation of vessel 8's sister ship vessel 7. Since the modulation appeared for a specific acoustic regime of a specific vessel design, it is likely a physical feature of the URN from these ships; however, its cause could not be identified.

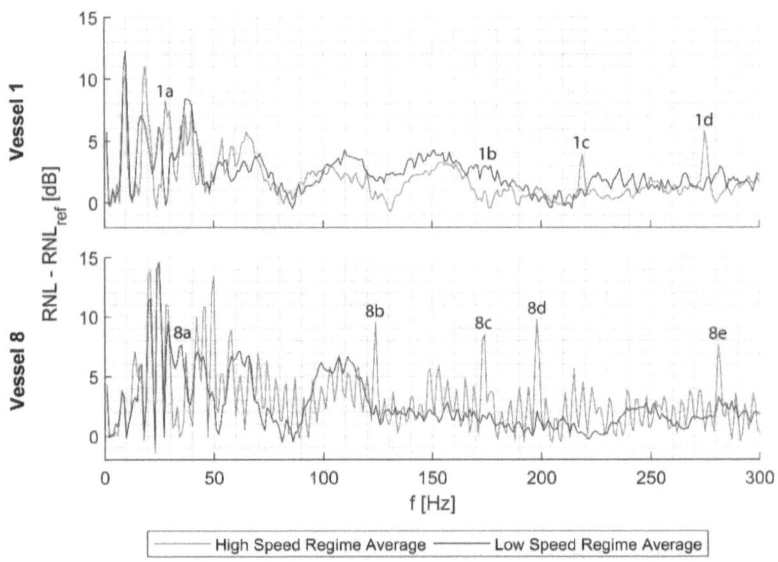

Figure 4.18: De-trended noise spectra in a linear x-axis scale of up to 300 Hz of (a) Vessel 1 and (b) Vessel 8 at low and high-speed regimes. Low frequency spectral features show distinct differences in shape and between different speeds and different vessel type. Noteworthy narrow-band features are identified with individual alphanumeric designations.

No direct observations of cavitation were made in the present work, and it is therefore impossible to definitively relate any spectral features to particular cavitation regimes. However, clear changes in peak location and structure in the low-frequency range between speed regimes corresponding to changes in broadband spectral features do strongly suggest a link between specific features and individual cavitation regimes. Of particular note was the presence of more pronounced tonal features in the low-frequency spectra of the high-speed acoustic regime, while sharp peaks were absent in the 100 to 300 Hz range for the louder low-speed acoustic regime. This finding may indicate that a reduction in tonal URN in this frequency range is indicative of a cavitation regime change that results in greater overall acoustic emission; however, confirmation of such a relationship requires further investigation. In addition to the broadband spectral features identified as being related to particular acoustic regimes, these narrow-band features could be useful for identifying acoustically undesirable propeller cavitation conditions in real time, allowing for active mitigation strategies.

4.6 Conclusions

Underwater radiated noise from eight coastal ferry vessels of various sizes, ages, and configurations was assessed using acoustic data collected from controlled, modified commercial operation. The relationship between radiated narrowband noise and the operating conditions of the vessels was investigated, with specific focus placed on the propeller and cavitation-related parameters. Speed was observed to be the most common primary determinant of narrow-band noise at the majority of frequencies. Noise was found to be anti-correlated with speed for some, but not all, ships that employed controllable-pitch propellers. The same trend was not observed for any ships using fixed-pitch propellers. For vessels that showed anti-correlation between speed and radiated noise, the quietest operation was seen close to or above their designed service speeds, suggesting that an optimal speed range for radiated noise performance may exist for these vessels. Linear regression analysis provided limited insight into the physical mechanism of this speed-noise relationship, but existing experimental work lends credence to the hypothesis that the increase in radiated noise at low speeds is the result of a change in propeller-induced cavitation regimes. Regardless of the cause, the existence of an optimal speed range for some vessels with controllable-pitch propellers is relevant from a regulatory perspective, since it implies that speed limits or slowdowns, which have been proposed or implemented on occasion for the purpose of reducing radiated noise from vessels, may result in increased noise from some vessels. Further research is needed to discover what design factors, in addition to controllable-pitch propellers, lead to vessels that generate increased radiated noise at low speeds.

For the pair of vessels that showed the strongest anti-correlation between speed and noise levels, a clear pair of acoustic regimes, associated with high- and low-speed operation, were observed. At low frequencies, tonal components were identified as unique to the acoustic regime associated with high-speed operation, potentially representing acoustic markers of a specific cavitation regime. Broadband features of the spectra associated with these acoustic regimes qualitatively resembled those of the radiated noise from tip vortex cavitation-dominated and pressure side cavitation-dominated propeller noise observed in previous studies, further supporting the hypothesis that this change in acoustic regimes was associated with a change in cavitation regimes. A transition between these cavitation regimes would account both for the increased noise at low speeds and the atypical relationship between radiated noise levels and cavitation index. This change in cavitation regimes is expected to be correlated with changes in propeller pitch, which was observed in the present study. While it might be expected that such a change in radiated noise regimes would occur for a large number of vessels with controllable pitch propellers, a means for predicting that transition a priori has yet to be identified. The present results suggest that further examination of the acoustic regimes of real ships could lead to both predictive empirical models that could be used in vessel design, as well as simple tools for real-time acoustic identification and mitigation of undesirable cavitation types.

Chapter 5: Conclusions

Cavitation-induced noise from ship propellers is a harmful pollutant in marine ecosystems, where sound is vital to fauna for sensing and communication. Mitigation of this noise pollution requires knowledge of the fundamental nature of cavitation-induced noise, strategies for modelling it at multiple levels of fidelity, and optimization strategies to reduce emitted sound. This book aimed to examine and evaluate noise modelling strategies with potential value, and to provide insight into future improvements to those strategies.

5.1   Summary of principal results

A discussion of propeller cavitation-induced noise from shipping as a critical pollutant in the world's oceans, as well as an overview of the state of modelling the emission of that noise, was presented in the first chapter of this book. The remainder of the book was divided into three main body chapters, each one a paper discussing modelling cavitation-induced noise with a different approach.

Chapter two discussed a combined experimental and numerical study investigation of cavitation-induced noise from a scale model controllable-pitch propeller. In the experiments, acoustic signals were acquired from ten different operating conditions that resulted in four district cavitation regimes. The experimental conditions were replicated in the simulations, which used the Unsteady Reynolds-Averaged Navier-Stokes equations in conjunction with an acoustic analogy based on the equation of Ffowcs-Williams and Hawkings. Increased noise levels were associated with particular regimes observed in the experiments; cavitation on the pressure side of blades, which was observed in under-loaded and reduced-pitch operating conditions, generated particularly high levels of noise. These results highlighted the need for increased attention to off-design operation of ships when considering underwater noise. The numerical simulations were successful in predicting cavitation inception, cavitation regimes, and, in most cases, the broadband features of the radiated noise induced by cavitation. On the other hand, the numerical results also revealed shortcomings in the modelling methodology used. First, Unsteady Reynolds-Averaged Navier-Stokes were confirmed to be ill-suited to reproducing cavitation that exists within vortex structures, confirming existing published findings. Second, noise levels were over-predicted by several orders of magnitude. Further investigation into the cause of the over-prediction of noise indicated that the deficiency was most likely related to the use of the Schnerr and Sauer cavitation model. The second finding is of particular interest in the area of modelling, since the Schnerr and Sauer model is widely used on account of the simplicity of its implementation.

Chapter three presented a proof of concept of a mapping procedure designed to allow efficient prediction of radiated cavitation-induced noise within the same framework used for the prediction of chemical emissions from internal combustion engines. The intended purpose of this framework was to facilitate intelligent operation optimization for ships. Using the present method, a map of radiated noise due to cavitation on the parameter space of propeller torque and propeller revolution rate was produced for the model-scale propeller studied in chapter three. The proof of concept relied on panel method simulations alongside semi-empirical noise models. The results demonstrate the potential of the procedure; however, the method is not yet generalizable.

Chapter four examined cavitation-induced noise using field measurement data from coastal passenger ferries in modified commercial operation. These vessels demonstrated different relationships between their operating speeds and radiated noise levels. Since vessel slowdowns have been proposed and trialed as a strategy for mitigating noise pollution in critical habitats for marine fauna, the paper focuses on the vessels that produce higher levels of noise at reduced speeds. For those vessels, the quietest operation was observed when the travel speed was close to the designed service speed. At reduced speeds, a transition to a different regime of sound production was observed via a change in the broad- and narrow-band spectral features. The spectral features were comparable to those observed in the experimental conditions that produced cavitation dominated by the tip vortex or pressure side regimes, supporting the hypothesis that the acoustic regime changes corresponded to a change in cavitation regime associated with a change in pitch. Vessels that are susceptible to similar changes in cavitation regime are unlikely to benefit from speed limits or slowdowns, so identifying these ships is valuable from a regulatory perspective.

5.2   Future work

Predicting cavitation-induced noise from marine propellers remains an area in need of development. If modelling strategies are to be adopted in the shipping industry for environmental protection, they need not only improved reliability, but also more clearly understood ranges of applicability. The present work highlights several shortcomings in modelling strategies at multiple levels of fidelity, all of which have the potential to be used in the field for design, control optimization, or regulatory purposes.

Methods for predicting cavitation-induced noise using computational fluid dynamics are still in active development. The results presented in chapter two suggest that an investigation into cavitation models themselves, in the context of cavitation-induced noise specifically, would provide a significant benefit to the field. While a limited number of these models have been applied to CFD-based prediction of cavitation-induced noise successfully, comparisons between models and investigations of their range of applicability are scarce in this context. A systematic study of cavitation mass transfer models for the purpose of acoustic analysis, accounting for the effects of the numerical framework, mesh, time domain, frequency range, hydrodynamic quantities, and scaling factors, would provide an invaluable baseline for future development in the area.

Reduced-order modelling, including techniques such as the mapping strategy presented in this work and the use of panel methods combined with semi-empirical noise models generally, is a promising approach to predicting cavitation-induced noise for the purpose of operation optimization. Unlike a tool such as URANS or LES, which are computationally expensive, it is feasible to run large numbers of panel method simulations in a relatively short timeframe, even without access to high-performance computing. This level of accessibility is beneficial for a technology aimed at environmental protection, as ease of adoption is critical. Panel methods are well-established; however, a great deal of room for innovation and advancement exists in the area of semi-empirical noise modelling. This **book** has shown that changes in cavitation regime are of primary importance to predicting radiated noise from marine propellers. In order for a semi-empirical noise model to be widely useful for the optimization of marine vessel operation, it must be able to account for a range of cavitation types commonly induced by the range of modern propellers, as transition between those regimes. This is no small task; it is necessary to perform both model-scale and full-scale experiments on multiple propeller types to create a semi-empirical noise model with a wide range of applicability.

One of the key challenges associated with predicting cavitation-induced noise from propellers is a lack of access to proprietary propeller geometries. Over the course of conducting the work presented in this **book**, it became clear that it is not uncommon for a vessel operator to have extremely limited information on the propellers used by their ships. The operator of the ferries studied in Chapter 4 did not have access to sufficient information on propeller geometry to perform even a simple reduced-order hydrodynamic simulation for any of these ships. A workaround for this issue may exist in the form of propeller surrogacy for modelling. Open-source, scalable propeller designs are commonplace. It may be possible to substitute an unknown propeller geometry for a visually similar open-source design for the purposes of reduced-order noise modelling; however, this concept has not been thoroughly investigated to date.

Appendix A: Turbulence Modelling in the Present URANS Simulations

Flow was modelled in Chapter 2 using the incompressible URANS equations, which are based on the decomposition of the instantaneous velocity field in to mean and fluctuating components. Mean components are time averaged over some interval T which must be significantly larger than the timescale of turbulent velocity fluctuations and smaller than the timescale of the evolution of the velocity mean field:

$$\bar{u}_i = \frac{1}{T} \int_t^{t+T} u_i dt. \tag{A1}$$

After Reynolds averaging the only fluctuating term remaining in the URANS equations is the Reynold's stress tensor $-\rho \overline{u_i' u_j'}$, which must be estimated by turbulence modelling. The URANS equations have the form:

$$\frac{\partial \bar{u}_i}{\partial x_i} = 0, \tag{A2}$$

$$\rho \frac{\partial \bar{u}_i}{\partial t} + \frac{\partial}{\partial x_j} \left(\rho \bar{u}_i \bar{u}_j + \rho \overline{u_i' u_j'} \right) = -\frac{\partial \bar{p}}{\partial x_i} + \frac{\partial}{\partial x_j} \left[\mu \left(\frac{\partial \bar{u}_i}{\partial x_j} + \frac{\partial \bar{u}_j}{\partial x_i} \right) \right]. \tag{A3}$$

The realizable k-ε two-equation turbulence model was used in the present simulations. The realizable k-ε model modifies the standard k-ε model in two ways; first the Reynolds stresses are replaced in the dissipation rate equation by a "source" term, and second a different eddy viscosity equation is used which accounts for the effects of mean rotation. The result is a model that behaves better for flows with high mean shear rates or large separation. The transport equations for turbulent kinetic energy k and dissipation rate ϵ become:

$$\frac{\partial k}{\partial t} + \frac{\partial}{\partial x_i} \left(k u_j \right) = \frac{\partial}{\partial x_i} \left[\left(\nu + \frac{\nu_T}{\sigma_k} \right) \frac{\partial k}{\partial x_j} \right] - \overline{u_i' u_j'} \frac{\partial u_i}{\partial x_j} - \epsilon, \tag{A4}$$

$$\frac{\partial \epsilon}{\partial t} + \frac{\partial}{\partial x_j} \left(\epsilon u_j \right) = \frac{\partial}{\partial x_j} \left[\left(\nu + \frac{\nu_T}{\sigma_\epsilon} \right) \frac{\partial \epsilon}{\partial x_j} \right] - C_1 S^* \epsilon - C_2 \frac{\epsilon^2}{k + \sqrt{\nu \epsilon}}, \tag{A5}$$

where ν_T is the turbulent viscosity, σ_k and σ_ϵ are the turbulent Prandtl numbers for k and ϵ respectively, S^* is the mean strain rate, and C_2 is a constant. C_1 is not a constant in the realizable model as it is in the standard k-ε model, but instead depends on the strain rate:

$$C_1 = max\left[0.43, \frac{\eta}{\eta + 5}\right], \quad \eta = S^2 \frac{k}{\epsilon}. \tag{A6}$$

The eddy viscosity model is:

$$\mu_T = \rho \frac{C_\mu k^2}{\epsilon}, \tag{A7}$$

$$C_\mu = \frac{1}{A_0 + A_s U^{(*)} \frac{k}{\epsilon}}. \tag{A8}$$

The non-constant terms A_s and $U^{(*)}$ capture the rotation and shear of the fluid element is three dimensions. Their formulations, as well as the remaining model coefficients, are given in [54].

Closure of the system of governing equations was obtained by calculation of the Reynold's stress tensor by the Boussinesq approximation:

$$-\overline{u_i' u_j'} = \nu_T \left(\frac{\partial \bar{u}_i}{\partial x_j} + \frac{\partial \bar{u}_j}{\partial x_i}\right) - \frac{2}{3} k \delta_{ij}. \tag{A9}$$

Appendix B: Implementation of the Panel Method

The panel method is a numerical BEM that uses discretized potential flow equations to solve for the distribution of pressure and velocity over the surface of a submerged body in a flow. Potential flow theory is valid in cases where the flow inviscid and irrotational; therefore, in cases where boundary layers may be assumed very thin, primarily classes of flows for which the Reynolds number is very high, potential flow theory is can be used to provide a simple approximation of the flow characteristics without the need to consider viscous effects.

Given a flow that is incompressible as well as inviscid and irrotational:

$$\nabla \times \vec{u}(x, y, z) = 0, \qquad (B1)$$

$$\nabla \cdot \vec{u}(x, y, z) = 0. \qquad (B2)$$

Since both the curl and divergence of velocity are zero, the flow velocity may be described in terms of a potential ϕ which satisfies Laplace's equation:

$$\nabla^2 \phi(x, y, z) = 0. \qquad (B3)$$

Additionally, the disturbance of any body moving at a velocity $\vec{v}(x, y, z)$ must go to zero far from that body, giving the boundary condition:

$$\lim_{r \to \infty} (\nabla \phi - \vec{v}) = 0. \qquad (B4)$$

The solution to Laplace's equation on some arbitrary domain can be determined by application of Green's identity, which provides the formulation used in ROTORYSICS [24], [72], [73], [84], following the development of the equations given in [85]. Consider potential flow within a volume V bounded by a disconnected surface S_0. Let S_0 be comprised of a boundary S_∞ about the entire flow region (which may be infinite) and a boundary S_B representing some object immersed in the flow, as shown in Figure B1.

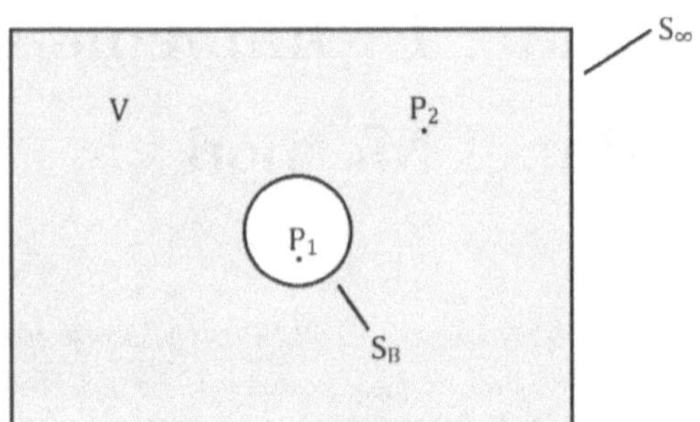

Figure B1: A hypothetical potential flow domain, including one point P_1 not on the domain and one point P_2 on the domain

Consider the continuously differentiable vector function:

$$\vec{F}(x,y,z) = \frac{1}{r(x,y,z)}\nabla\phi(x,y,z) - \phi(x,y,z)\nabla\frac{1}{r(x,y,z)} \, . \qquad (B5)$$

Here $r(x,y,z)$ is the distance from some point P to point (x,y,z). Applying Gauss' divergence theorem to this function on V gives:

$$\oiint_S \vec{F}(x,y,z)\cdot\vec{n}\,dS = \iiint_V \nabla\cdot\vec{F}(x,y,z)\,dV, \qquad (B6)$$

$$\oiint_{S_0}\left(\frac{1}{r}\nabla\phi - \phi\nabla\frac{1}{r}\right)\cdot\vec{n}\,dS = \iiint_V\left(\frac{1}{r}\nabla^2\phi - \phi\nabla^2\frac{1}{r}\right)dV. \qquad (B7)$$

In the case that P is outside V, such as P_1 in Figure B, both ϕ and $1/r$ satisfy Laplace's equation and Eqn. (B7) becomes:

$$\oiint_S\left(\frac{1}{r}\nabla\phi - \phi\nabla\frac{1}{r}\right)\cdot\vec{n}\,dS_0 = 0. \qquad (B8)$$

For points of interest outside V (generally inside submerged bodies) the potential is replaced by an "internal potential" ϕ_i which is a mathematical construction facilitating conformal mapping rather than a description of physical flow. The equation is also valid integrating only on the boundary of a submerged object:

$$\oiint_{S_B}\left(\frac{1}{r}\nabla\phi_i - \phi_i\nabla\frac{1}{r}\right)\cdot\vec{n}\,dS_0 = 0. \qquad (B9)$$

If P is on V, as is the case for P_2 in Figure B1, then $1/r$ does not satisfy Laplace's equation. In this case, consider a modified domain which excludes a spherical region of radius ϵ with a surface S_ϵ about point P:

$$\oiint_{S+S_\epsilon} \left(\frac{1}{r}\nabla\phi - \phi\nabla\frac{1}{r} \right) \cdot \vec{n}\, dS = \oiint_S \left(\frac{1}{r}\nabla\phi - \phi\nabla\frac{1}{r} \right) \cdot \vec{n}\, dS - \oiint_{S_\epsilon} \left(\frac{1}{r}\frac{\partial\phi}{\partial r} + \frac{\phi}{r^2} \right) dS = 0. \quad (B10)$$

Since the region is arbitrary, limit as ϵ goes to zero may be taken, which causes the derivative terms to vanish and gives:

$$-\lim_{\epsilon\to 0} \oiint_{S_\epsilon} \left(\frac{1}{r}\frac{\partial\phi}{\partial r} + \frac{\phi}{r^2} \right) dS = -\lim_{\epsilon\to 0} \oiint_{S_\epsilon} \left(\frac{\phi}{r^2} \right) dS = -4\pi\phi(P). \quad (B11)$$

Equation B10 then becomes:

$$\phi(P) = \frac{1}{4\pi} \oiint_S \left(\frac{1}{r}\nabla\phi - \phi\nabla\frac{1}{r} \right) \cdot \vec{n}\, dS. \quad (B12)$$

Thus, the velocity potential at any point in the flow can be computed using only information from the boundaries of the domain. This solution always satisfies the boundary condition (equation B4). There is one further case of note; if P is on the boundary S (and within V) then it becomes necessary to exclude a hemispherical region about P from the domain during the aforementioned treatment, and equation B12 becomes:

$$\phi(P) = \frac{1}{2\pi} \oiint_S \left(\frac{1}{r}\nabla\phi - \phi\nabla\frac{1}{r} \right) \cdot \vec{n}\, dS. \quad (B13)$$

Combining equations B9 and B13 and applying them to a domain with N submerged bodies gives:

$$\phi(P) = \frac{1}{4\pi} \sum_{1}^{N} \oiint_{S_{B_N}} \left(\frac{1}{r}\nabla(\phi - \phi_i) - (\phi - \phi_i)\nabla\frac{1}{r} \right) \cdot \vec{n}\, dS,$$

$$+ \frac{1}{4\pi} \oiint_{S_\infty} \left(\frac{1}{r}\nabla\phi - \phi\nabla\frac{1}{r} \right) \cdot \vec{n}\, dS. \quad (B14)$$

In order to make these generalized solutions for potential flow useful for application to lifting bodies, they are usually expressed in terms of sources and doublets along the boundary surfaces. Source and doublet strength may be expressed in terms of ϕ and ϕ_i:

$$-\mu = \phi - \phi_i, \quad (B15)$$

$$-\sigma = \frac{\partial \phi}{\partial n} - \frac{\partial \phi_i}{\partial n}. \tag{B16}$$

Where μ is doublet strength and σ is source strength. Note that source is conventionally denoted by the symbol σ, but that symbol is also used elsewhere in this book to refer to cavitation number. The operator $\frac{\partial}{\partial n}$ is equivalent to $\vec{n} \cdot \nabla$ here. Thus, equation B14 may be expressed, finally, as:

$$\phi(P) = \frac{-1}{4\pi} \sum_1^N \oiint_{S_{B_N}} \left(\sigma \frac{1}{r} - \mu \frac{\partial}{\partial n} \left(\frac{1}{r} \right) \right) dS + \phi_\infty(P). \tag{B17}$$

Where $\phi_\infty(P)$ depends on the reference frame. For example, in the case of a body moving through otherwise stationary fluid, $\phi_\infty(P)$ is constant (and may be chosen arbitrarily). For flow over a single, fixed body:

$$\phi_\infty(P) = u_\infty x + v_\infty y + w_\infty z. \tag{B18}$$

Flow over a lifting body with a sharp trailing edge creates a shear layer in the wake, causing a discontinuity in the potential. To handle this discontinuity in potential solvers (including ROTORYSICS), the Kutta condition is commonly applied. The Kutta condition asserts that circulation about a lifting body with a sharp trailing edge will be sufficiently large such that a stagnation point exists at the trailing edge. In practise, the result is that the shear layer behind lifting bodies is replaced in a panel method model by a thin body that extends indefinitely (i.e. many chord lengths) downstream from the trailing edge. This "wake body" is sufficiently thin such that the internal potential is continuous with the potential at the boundaries:

$$\frac{\partial \phi}{\partial n} = \frac{\partial \phi_i}{\partial n}. \tag{B19}$$

As a result, the source strength is zero on wake bodies. The potential for a flow over a single lifting body is therefore given by:

$$\phi(P) = \frac{-1}{4\pi} \oiint_{S_B} \left(\sigma \frac{1}{r} - \mu \frac{\partial}{\partial n} \left(\frac{1}{r} \right) \right) dS + \frac{1}{4\pi} \oiint_{S_W} \left(\mu \frac{\partial}{\partial n} \left(\frac{1}{r} \right) \right) dS + \phi_\infty(P). \tag{B20}$$

The distribution of sources and doublets in equation B17 must be chosen to suit the physics of the problem. In a panel method, source-doublet pairs are positioned at the centroids of polygonal panels that make up a surface mesh approximating the geometry of submerged bodies. Choosing the "internal potential" of bodies to be constant yields a zero value for potential at each point within a body. This treatment gives the Dirichlet boundary condition for the BEM scheme. This formulation is only valid for a unit source strength of:

$$\sigma = \vec{n} \cdot \vec{V_\infty}, \qquad (B21)$$

where $\vec{V_\infty}$ is the total kinematic velocity due to motion of the body. The analytical expressions for the influence of a polygonal surface panel with constant doubled and source strengths on the velocity potential at any point in space are:

$$\phi_{\text{doublet}} = \frac{z\mu}{4\pi} \iint\limits_{\text{panel}} [(x - \sigma)^2 + (y - \mu)^2 + z^2]^{-3/2} d\sigma d\mu, \qquad (B22)$$

$$\phi_{\text{source}} = -\int\limits_{z}^{\infty} z d\phi_{\text{doublet}} - z\phi_{\text{doublet}}. \qquad (B23)$$

For a quadrilateral panel these equations can be represented in numeric form, as shown in [84].

For a given point P within a solid body with N surface panels, the numerical form of equation 2.17 may be written as:

$$\sum_{k=1}^{N} C_k \mu_k = -\sum_{l=1}^{N_W} C_{W_l} \mu_l - \sum_{k=1}^{N} B_k \sigma_k, \qquad (B24)$$

where C_k and B_k are the coefficients of influence of panel k on point P, found from the numeric forms of Eqns. B22 and B23. The subscript W here indicates a wake panel. One such equation is formulated for each collocation point, i.e. a point located just inside a body at the centroid of each panel. The values of σ_k are known from equation 4.1 (also evaluated at the centroid of each panel), and the doublet strengths of wake panels can be computed iteratively at each time step from the Kutta condition, so the system can be solved for μ_k.

In the present work, multibody interactions are used, as the propeller and hub are a separate body from the "pod" (the conical section upstream of the propeller in Figure 2.3). One system was created for each body separately, and multi-body interactions were implemented iteratively. The resultant matrix equation (for each time step) took the recursive form [86]:

$$\left([C^1]_{(I_1,I_1)}[\mu^1]_{I_1}\right)^{\mathrm{m}} = \ [B^1]_{(I_1,I_1)}[\sigma^1]_{I_1} - [C_W^1]_{(I_1,J_1)}[\mu_W^1]_{J_1} \cdots$$
$$-\left([C^{2\to1}]_{(I_1,I_2)}[\sigma^2]_{I_2}\right)^{\mathrm{m}-1} \cdots$$
$$-[B^{2\to1}]_{(I_1,I_2)}[\sigma^2]_{I_2} - [C_W^{2\to1}]_{(I_1,J_2)}[\mu_W^2]_{J_2} \cdots \qquad (B25)$$
$$-\left([C^{N\to1}]_{(I_1,I_2)}[\sigma^N]_{I_2}\right)^{\mathrm{m}-1} \cdots$$
$$-[B^{N\to1}]_{(I_1,I_N)}[\sigma^N]_{I_N} - [C_W^{N\to1}]_{(I_1,J_N)}[\mu_W^N]_{J_N}$$

Where:

- $[X]_{(i,j)}$ is an i by j matrix of elements of type X
- $[C^k]$ is the coefficient of influence matrix of the doublets of body k on the panels of body k
- $[B^k]$ is the coefficient of influence matrix of the sources of body k on the panels of body k
- $[C_W^k]$ is the coefficient of influence matrix of the doublets of wake body k on the panels of body k
- $[C^{h\to k}]$ is the coefficient of influence matrix of the sources of body h on the panels of body k
- $[B^{h\to k}]$ is the coefficient of influence matrix of the sources of body h on the panels of body k
- $[C_W^{h\to k}]$ is the coefficient of influence matrix of the doublets of wake body h on the panels of body k
- I_k is the number of panels on body k
- J_k is the number of panels on wake body k
- m is the iteration number

A maximum of five iterations of Eqn. (B25) were allowed in order to reach convergence at each time step in the present work. Convergence was determined by the relative change in the induced velocity [86]. For the first iteration, the doublet strengths on each of these surfaces are calculated without including the influence of doublets on other surfaces.

Once multibody interactions were solved, potentials were corrected for the influence of other bodies. Simple numerical differentiation according to the unsteady Bernoulli equation then gave the pressures and velocities for each panel on the submerged bodies, which could in turn be integrated to obtain body forces.